FLOYD EARL SMITH

Runaway

*How the Earth "Tipped"
into Runaway Climate Change*

RUNAWAY COMMUNICATIONS
London, UK

Published by Runaway Communications
8 Fairholme Road, London W14 9JX

Information on this title: www.getridofwarming.org

© Floyd Earl Smith, 2009

This book is in copyright. Subject to statutory exception and to the provisions of relevant collective licensing agreements, no reproduction of any part may take place without the written permission of Floyd Earl Smith.

First Edition

ISBN-13: 978-0-9553300-3-2
ISBN-10: 0-9553300-3-3

To James and Veronica
With love and hope

Table of Contents

Preface .... **vii**

Introduction .... **ix**
    When no news is better than bad news x
    What hope is there? xi

Chapter 1. Environmental Megatrends **1**
    Climate cycles 3
    Population growth 6
    Deforestation 9
    Lack of fresh water 11
    Increasing vulnerability to rising seas 13
    "Western" lifestyles 16
    Extinctions and declines in numbers 18
    Megatrends and climate change 20

Chapter 2. Climate Change 101 .... **23**
    Why can't we all just get along? 25
    "Commonsense" evidence 26
    "Scientific" evidence 28
    The causes of climate change 30
    How emissions hang around 33
    The agony and the enthalpy 34
    Kyoto vs. climate change 35

Chapter 3. The 2007 Consensus **41**
    What is the 2007 Consensus? 42
    Gore's inconvenient truths 45
    The broken IPCC process 46
    Hansen's Alternative Scenario 51
    A none too Stern review 52
    More false hope 53
    A Gaia future? 55

Chapter 4. Why Emissions are Bound to Rise **57**
    Whence emissions 58
    Why emissions matter 60
    Emissions and the 2007 Consensus 61
    How growth is tied to emissions 63
    Why emissions won't easily drop 67
    Why CO_2 levels won't drop at all 69
    S curves and tipping points 70
    A more realistic curve for emissions 72
    IMechE and others sound off 75
    Can emissions really keep growing? 78

Chapter 5. Emissions Growth Worldwide **81**
    The UK and Europe 82
    Emissions and the US 83
    China and the developing world 85
    Russia and Canada 90

Chapter 6. How Carbon Sinks Have Already Tipped ... 93
- What are the carbon sinks? ... 95
- The North Pole's ice cover ... 99
- The direct effect ... 101
- The yedoma erupts ... 102
- Greenery destruction ... 105
- Other carbon sinks ... 107

Chapter 7. A Model for Runaway Climate Change ... 111
- How the dominoes work ... 111
- What a model does for us ... 114
- Filling in the model ... 115
- The permanence of effects ... 123
- What is the total impact? ... 124

Chapter 8. How the World Will Worsen to 2050 ... 127
- The end of the population boom? ... 128
- Scarce water and Peak Food ... 131
- Air pollution and climate change ... 136
- Extinctions ... 137
- The direct effects of climate change ... 138
- The human response to climate change ... 139
- Toward Hot Earth? ... 141

Chapter 9. Fighting Climate Change ... 143
- Too late by half? ... 145
- A framework for saving the Earth ... 146
- Goal 1. Set out strong founding principles ... 148
- Goal 2. Get the science right ... 153
- Goal 3. Dig slower ... 155
- Goal 4. Zero tolerance for CO_2: Power ... 158
- Goal 5. Zero tolerance for CO_2: Transport ... 162
- Goal 6. Back to 280ppm and pH 8.2 ... 165
- Goal 7. Lower the temperature ... 167
- Mechanisms for encouraging compliance ... 170

Chapter 10. How Nations Can Respond ... 173
- A template for nations ... 174
- Case study: United Kingdom ... 180
- Case study: United States ... 186

Appendix A. FAQ ... 195

Appendix B. Rallying Cries ... 207

Appendix C. Selected Resources ... 209

Appendix D. Winners and Losers ... 217

Appendix E. Discussion questions ... 227

Index ... 231

List of Figures

Figure 3-1. Four IPCC emissions scenarios...................................47
Figure 3-2. Four IPCC emissions scenarios plus 3% growth........49
Figure 4-1. US emissions by source 58
Figure 4-2. US emissions, electricity divided among users 59
Figure 4-3. Greenhouse gas emissions since industrialisation ...65
Figure 4-4. Greenhouse gas emissions, converging by 2050.......73
Figure 6-1. North Pole sea ice extent for recent years................100
Figure 7-1. Climate change dominoes model 121

Preface

> "A new scientific truth does not triumph by convincing its opponents and making them see the light, but rather because its opponents eventually die, and a new generation grows up that is familiar with it." Max Planck

Decades after it was first raised as a concern, people around the world are finally aware of global warming and climate change. But the story as currently understood contains many flaws, and is disconnected from the longer-term trends that give climate change its awful speed (in relative terms) and power. My intent is to give you a more accurate, broader, deeper understanding that will equip you for the difficult decades ahead.

I'm especially hopeful that this book will come to the attention of young people between the ages of 16 and 30. For those of you who are young adults, I expect that this book will help you to better understand the fast-changing world in which you'll live out your adult lives and just what you, in turn, may hand on to your children and grandchildren.

It's particularly directed to you because most of us fix our major ideas about the world when we're young. We only have "beginner's mind", as the Buddhists put it, while we are, in fact, beginners. After that, for better or worse, we act within a relatively fixed mental framework.

An example of this is the car – a pertinent example for a book about climate change. Henry Ford offered the affordable, mass-produced Model T shortly after 1900. But American car culture – a huge rush of books, magazines, songs, new dating rituals and new working patterns – didn't come on strong until the 1950s, when most American families had at least one car, and any young man who wasn't a complete nerd knew how to gap a spark plug.

A more recent example is the mobile phone. I had one in the late 1990s, in my mid-30s, and as such was an early adopter. I was also, at least for an American, a fairly enthusiastic user of text messaging. But one day I handed the phone to my then ten-year-old daughter so she could send a text.

Her fingers danced, smoke rose from the phone's keys, and the phone flew back into my hands. The task was so easy for her that I realised the mobile phone was, in her hands, a different instrument than in mine. She was growing up with it, and I hadn't.

Unfortunately, the same is true for climate change. Though discussed and debated since the late 1970s, it only became a major public concern in 2006 and 2007, largely due to the efforts of former US Vice President Al Gore. Only people who were still young and impressionable when Gore made the world aware of the problems – and as follow-on books like this one come out – are likely to be able to really understand what climate change will mean to the Earth, and to themselves, in the decades ahead.

Older people might have trouble taking the worries outlined by Gore and in books like this one entirely seriously, simply because their ideas are already fixed and any effect on them personally is low. I've seen this lack of engagement among my own friends and acquaintances – with a few honourable exceptions – and the same effect has been reported by university professors who have a mix of younger and older students in their classes and lectures on climate change. Only the younger students "get it".

This disparity is unfortunate, because in this book I propose a further development that will be even harder to absorb. I assert that runaway climate change – which means out of control and self-sustaining changes in the environment, leading to further warming – which Gore and many others have claimed can be avoided if very strong action is taken by mid-century, is actually already in progress, and has been since roughly the early 1980s.

The effects of not just climate change, but runaway climate change, are already in evidence; barring a near-miracle, they will become quite dramatic in the next few decades, and progressively reshape our world throughout the century. There will not only be more and more heat and intense weather, but less and less fresh water for growing food, less arable land available on which to grow it, shifting climactic zones, massive losses of wildlife, and further strains as the effort to fight climate change shifts into high gear.

Which means that those of us over 30 have left those of you who are younger – our literal and metaphorical children – a very rough road indeed. I can only hope that this book can help to begin to light a tenable way forward.

Introduction

> "We are facing a climate crisis that is larger and harder to deal with than any of us thought." – Chris Field, IPCC, 2009

Climate change is not as big a threat as most people think.

It's a worse threat. Much worse.

In fact, it's worse than just about anyone realises.

Everyone who understands the science behind the situation – or who is willing to listen to mainstream scientific opinion – "knows" two things: that global warming will increase significantly if we don't cut emissions; and that nature is in danger of "tipping" into runaway warming if the cuts don't come within a few decades.

This "knowledge" is wrong in both of its particulars. First, emissions are much higher than projected even a few years ago, and all the momentum is on the side of further increases. So cutting emissions below the trend of the previous, too-low estimates is just about impossible.

Emissions projections were prominent in Al Gore's book and movie, *An Inconvenient Truth*, and in the 2007 report of the Intergovernmental Panel on Climate Change, the world's leading body for assessing the results and implications of climate change research. But the projections in the Gore works and – less excusably – in the 2007 IPCC reports were based on dramatic underestimates for China's emissions, as well as on 1990s worldwide emissions growth rates of just over 1% a year. The truth is that China's emissions in particular were much higher than realised, almost at US totals in 2007, then surpassing the US – to become the world's largest – in 2008. And since the year 2000, not unrelatedly, worldwide emissions have grown at an average rate of more than 3% a year.

The IPCC has also created and widely promulgated Scenarios that specify an absence of explicit control efforts – yet still somehow project even these too-low estimates dropping spontaneously, for no good reason, at varying points in this century. These Scenarios – and in particular a famous and widely reproduced graph of them, included in this book as Figure 3-1 – may be one of the most consequential pieces of scientific misinformation in history.

Though we may have a brief reduction in emissions growth rates for the current recession, a similar rate of growth is expected to resume, and then continue for decades to come. For the next few decades this growth is a simple and direct effect of more than 2 billion people – and growing steadily – in China, India and elsewhere moving from relative poverty to European or American-type lifestyles, with the cars, homes, meat and processed food consumption, air travel and "stuff" – plus the carbon emissions – that go with them. And there are billions more behind them, waiting to get on the escalator to Western living standards – and Western levels of emissions and deforestation.

This is a genie that will come all the way out of the bottle unless it's stoppered very harshly indeed. So worldwide emissions will continue to rise sharply in the coming decades, short of extensive and lasting war, natural disaster and/or economic disaster.

Along with emissions being worse than previously thought, I assert that we're also past the point of preventing tipping into runaway climate change. The first major carbon sink, the North Pole's ice cover, has already tipped – a process that's gathering momentum today. Its tipping point was reached and surpassed in the early 1980s, as I'll explain later in this book. Other carbon sinks are close behind.

So the news is bad in stereo – we aren't controlling our emissions, and even if we were to begin to do so now, we have already touched off the feared self-sustaining warming spiral in nature.

When no news is better than bad news

These conclusions are supported by recent news. Despite a great deal of talk about reducing emissions, they are in fact rising sharply. Among the countries that committed to reducing emissions under the Kyoto Protocol, many are reporting failure to meet their commitments, rather than what's needed: success, and a willingness to commit to dramatic new reductions in upcoming negotiations.

Also new is the accelerated loss of summertime surface ice at the North Pole, ice which is now on track to disappear half a century sooner than was anticipated just a few years ago. The summer of 2007 saw the size of the North Pole's ice cover shrink to a record low; the summer of 2008 saw the Northwest and Northern passages open, making the Pole circumnavigable for the first time in history. Going into the summer of 2009, the ice is thinner and newer than ever. Scientists now predict a North Pole completely free of ice in summertime by about 2030.

It's now clear that other changes predicted for the second half of this century – worsening weather that, among other problems, reduces food production; methane releases from land and sea that accelerate warming; temperature increases of more than 2°C; and sea level rises of up to a metre – are likely to come forward and happen in the decades between now and mid-century.

These disastrous changes will occur on top of, and exacerbate, other unfavourable environmental trends. Population growth and lack of fresh water are two among several problems that will increase the pressure on humanity. And, as I'll describe here, it's already too late to stop climate change from becoming runaway – not that we were ever likely to try very hard anyway, at least in the short term, because of the cost and disruption required.

Both parts of this deviation from received wisdom matter greatly: the runaway nature of humanity's emissions and deforestation, and the runaway climate change in nature. They both matter because the warming caused by runaway emissions and the warming caused by runaway changes in nature are additive to each other.

Climate change will accelerate, with consequences we can only outline: temperatures rising even faster than the current change of .25°C per decade – that's 2.5°C per century; ice disappearing; seas rising two metres per century or more for many centuries to come; CO_2 levels rising in the atmosphere and seas. As these changes continue – and, heaven forfend, accelerate – we can say goodbye to more and more of the species that populate today's natural environment, and perhaps to civilisation as well.

Where will it all end? Further research is needed, but an era like the Jurassic, but without the same richness or diversity of life, is a good first approximation. We are likely to move to a world with temperatures more than 10°C (18°F) higher than today, and therefore little ice outside refrigerators; a third or more of land area turned to desert; sea levels roughly 80m higher; seas and atmosphere saturated with CO_2; today's species mostly extinct, and a wait of millions of years for new forms of life to evolve to revel in the new environment and, over tens of millions of additional years, to suck up the excess CO_2. I call this near-future version of our planet "Hot Earth".

That's if our decimated, embittered descendants don't try their hand at planetary climate engineering and cobble together some other situation – possibly a better one, but also possibly one even worse than a Hot Earth-type world.

What hope is there?

It's far too late for business as usual. The world will change fast. Those of us over 40 – in particular, we "baby boomers", born between 1945 and 1960, who largely run things today – will mostly pass away before things get too bad. The designation some have made of our grandparents as "the greatest generation" will certainly go unchallenged by us.

Those of you who are young people today, on the other hand, will live your lives in a world filling with disaster and conflict. And you will do well indeed not to pass on an even worse situation to your own children and grandchildren.

What can be done? If the entire world was agreed on the problem and dedicated to finding a solution, quite a bit. A really focussed effort to cut emissions and deforestation – imagine a world with ever fewer cars, declining meat consumption and no more passenger planes – might buy decades of delay.

It's even possible that the changes in nature could then be reversed by cooling the planet, perhaps by putting particulates in the air to reflect sunlight, then seeding algae in the seas to soak up excess CO_2 in the seas and air. All carefully targeted to bring the world back to its pre-warming equilibrium and maintain it there.

But this would take massive scientific and technical advances and unprecedented international cooperation. The advances may or may not come, but the concerted action needed is even more unlikely than it might at first seem.

Why? Because some countries will benefit from climate change, at least in its early stages. Russia and Canada will be climate change's largest beneficiaries as the oil- and gas-rich Arctic – which each has recently claimed a huge chunk of – opens up, and vast steppes and plains several times the size of Europe, currently unproductive, become grazing and farmland. (In a fast-approaching world where oil, gas, meat and grain are all likely to be far more valuable than today.)

Russia in particular has already acted with foresight and determination to assert its rights in a warming Arctic, and may well resist concerted action to reverse climate change with some or all the considerable means at her disposal.

Not that she's likely to need to. The US and China, by far the biggest emitters, are on track to lose large parts of their countries to desertification. (As I write this, California's forests are aflame, again, and the sands continue their inexorable advance towards Beijing – which already suffers greatly from sandstorms each spring.) Yet in the crucial first eight years of this century, the US and China have between them slowed international action on climate change to a crawl, and are still hardly leading the charge to significantly reduce their respective emissions.

If you think great powers aren't willing to imperil the world in pursuit of their perceived national interests, World War I, World War

II and forty years of nuclear brinksmanship in the Cold War are good examples of the opposite.

All of these realities and trends are in opposition to the current, far more optimistic consensus on climate change. (Which is itself considered wildly pessimistic, even a hoax, by many.) In this book I describe the trends that are on track to put hundreds of millions of people in peril even before considering climate change; how we know that climate change is occurring; how a rough (and overoptimistic) agreement on the extent of the problem, which I call the "2007 Consensus", was reached; how emissions and changes in nature have already reached runaway level, contrary to the conclusions embodied in the 2007 Consensus; and what we might hope to do about the dire situation that, all unawares, we have already been in for some time.

But before we can even begin to address these problems, we need to move, very quickly, from our previous understanding to the new one described herein. Such paradigm shifts, as they're called, generally take decades. Their initial proponents often pass away, as Planck described, before the new approach is accepted.

In the case of Charles Darwin's revolutionary theory of evolution, resistance began immediately, and continues today, more than 150 years later.

And consider the strange case of Dr Ignaz Semmelweis, who in the mid-1800s tried to convince doctors to wash their hands before delivering babies, so their mothers wouldn't develop infections and die just after childbirth. Semmelweis was driven insane by resistance and by his failure, over decades, to convince others of his views.

We don't have decades to take action of unprecedented boldness against climate change, and I for one hope to hold onto my sanity. So here's hoping the paradigm shift we need can happen more quickly, and more smoothly, than some others have done.

Chapter 1. Environmental Megatrends

> In this chapter
> Drivers, well established before climate change, of upcoming environmental problems:
> 1. Long-term cycles
> 2. Population growth
> 3. Deforestation
> 4. Overuse of water
> 5. Clustering on coasts
> 6. "Western" lifestyles
> 7. Extinctions

Mankind has become the dominant species on Earth, with rapidly increasing numbers, growing wealth and, up to now, improving health. Human beings are at the top of every food chain on Earth, most of which are managed by us – in some cases, right down to the genetic makeup of the species involved. Yet we are also in unprecedented danger.

This is not an accident. In fact, it's the result of long-term trends, several of which also lay the groundwork for climate change and the problems it will cause.

I call these long-term trends "environmental megatrends" to differentiate them from shorter-term trends in the environment and in society. The term "megatrends" was popularised by the book *Megatrends* by John Naisbitt (1986).

These environmental megatrends have combined to place humanity in a difficult, even dangerous, position – even before climate change and runaway climate change are taken into consideration. (Al Gore didn't call his 1993 environmental bestseller *Earth in the Balance* for nothing.) If climate change quickly causes massive problems, it will in large part be due to the fact that humanity was already headed for trouble before they arrived.

Environmental megatrends are existential – they relate to the conditions of current human existence. In distinction to the trends described in Naisbitt's book, environmental megatrends tend to operate at a scale of centuries, millennia or more. However, we have it in our power to accelerate or slow them; and to either succumb to them, or manage them more or less successfully over the next few decades.

These megatrends are like a set of multi-dimensional Russian dolls, each nested within and shaped by the others. Understanding their deep roots and interrelatedness is crucial to the success or failure of efforts we make to slow, stop and reverse global warming. (Sticking plasters are unlikely to work.)

Here are seven environmental megatrends which lay the groundwork for climate change and runaway climate change:

1. **Climate cycles**. Humanity's evolution and the development of civilisation have taken place within a series of changes to the climate, some catastrophic, others gradual. (Climate change due to a volcanic eruption may have reduced human population to a few thousand souls about 70,000 years ago.) These changes are most vividly marked by Ice Ages, when glaciers cover about a quarter of the Earth. We are currently in an interglacial period, referred to as the Holocene, that began about 11,000 years ago.
2. **Population growth**. Aided by the recent benign period in the climate and the development first of agriculture, then of industry, human population is on a solid curve of exponentially increasing growth, currently at 6.1 billion people and adding nearly a billion people per decade.
3. **Deforestation**. Over a period of millennia, accelerating with population growth, people have already cleared roughly half the world's forests – with monks, for instance, clearing much of Europe's former forests in the Middle Ages. Deforestation reduces the absorption of greenhouse gases emitted by people and the production of oxygen people need for life. Agricultural, industrial and transportation needs drive deforestation ever faster; little of our current forest cover is on track to survive.
4. **Lack of fresh water for crops**. Agriculture in a given region succeeds or fails largely on the availability and sustainable use of fresh water, determining the success or collapse of civilisations. As human numbers grow, and as water-conserving forests, glaciers and aquifers are destroyed, water crises are increasing. This trend, by itself a disaster unfolding worldwide, is worsened by climate change and amplifies its damaging effects in turn.
5. **Clustering on coasts**. Rising seas are a trailing indicator of climate change – we may see a global temperature increase of 5°C (9°F) before we experience sea level rises of much more than a metre. Yet we have crowded onto coastal lands – with about half the world's population living within 100 miles of the coast – so that even small sea level rises and accompanying increases in storm intensity and damage are extremely threatening.

6. **"Western" lifestyles**. Industrialised or "Western" lifestyles, represented by automobiles, jet planes, mechanised farming and cattle fattening on feedlots, are wildly attractive, freeing people from servitude to the land. But Western lifestyles generate game-changing quantities of pollutants, with greenhouse gases only the latest threat. The share of all people living Western lifestyles is at about a billion people – one-sixth of all people alive today – and growing rapidly as China and India, in particular, industrialise.
7. **Extinctions and reductions in numbers of wildlife**. We tend to apply an "either/or" mentality and focus on absolute extinctions – which are soaring – but reductions in numbers short of extinction are almost as important. Land and seas that once teemed with plants and wildlife have been decimated by human pressure as our numbers have increased. Entire ecosystems and interrelated webs of species are due to be damaged and then extinguished as climate change escalates. Fear of the effects on humanity alone, as food sources disappear and air and water quality worsen, should already be generating a massive response.

Following is a brief description of the extraordinary confluence of trends that has led us to a point of crisis even before climate change is considered. Which makes it very easy indeed for climate change – and, most particularly, runaway climate change – to send us hurtling toward disaster.

Climate cycles

The Earth's climate has changed a great deal since the planet first cooled enough to host liquid water about 4 billion years ago. Radiation from the sun has varied, gradually decreasing overall by a total of about 30%. (If we could only wait for another billion years or so, our global warming problem might be solved.)

One big change that has affected the climate is the separation of the planet's landmasses, when the giant supercontinent called Pangaea split up into today's continents beginning about 200 million years ago.

In Africa, the last common ancestor between humans and our closest relatives, chimpanzees and bonobos, lived about 5 to 7 million years ago. The genus Homo is estimated to be about 2 to 3 million years old, its emergence marked by the first evidence of stone tools. Homo means "the same as" – the same as us.

Our specific species, Homo Sapiens – Sapiens meaning knowledgeable – diverged from Neanderthals, who had bigger brains than us, and other close cousins only about 250,000 years ago, then outcompeted or more directly demised all the others.

Humans' genetic inheritance, plus the corresponding cultural evolution that have together caused us to diverge from our nearest neighbours on the tree of life, are what have allowed humans to "fill the Earth, and subdue it", as Genesis puts it.

To understand how the Earth's climate cycles relate to where we are today, it's good to know what past extremes have been and what has occurred in the more recent past.

We have only a general idea of what Earth's climate and CO_2 levels have been during most of the planet's history. Some minerals and some fossils only form at certain temperatures and pressures, so we have some idea from these sources. Even when we find useful minerals or fossils, dating techniques for periods millions or hundreds of millions ago usually have significant margins of error.

Levels of CO_2 in the atmosphere and global temperatures have been closely related throughout this period. However, it's not always clear whether a rise in CO_2 preceded a rise in temperature, the opposite, or if they rose or fell together.

One estimate is that there has been permanent ice on Earth's surface for periods totalling only about one fifth of Earth's history. Yet apparently Earth has also, in at least one instance about 600 million years ago, been "Snowball Earth", with low atmospheric CO_2 levels and completely covered with ice. The high reflectivity of ice kept the Earth very cold until a rare cluster of volcano eruptions pumped enough CO_2 into the atmosphere to restore the greenhouse effect and melt most of the ice.

Showing the effects that life can have on climate, Snowball Earth conditions may have been caused by living creatures absorbing carbon into their bodies, then being interred at the bottom of the sea. The gradual reduction of carbon, and thence CO_2, from the biosphere caused temperatures to crash and glaciers to advance. Only the volcanic outbreak pumped enough CO_2 into the atmosphere to reverse glaciation. (This is an instance where it's clear additional CO_2 preceded, and seemingly caused, warming.)

Apparently there have been no Snowball Earth conditions since worms evolved about 500 million years ago. Worms stir up sediments, freeing carbon from internment and preserving CO_2 levels in the near-surface environment.

At the other extreme from Snowball Earth are what I call Jurassic Park conditions – hot periods during which the average global temperature was about 10°C above today's and greenhouse gas levels were about four times higher than recently. These periods are largely free of surface ice on Earth. The Jurassic period itself began about 200 million years ago, just as Pangaea was breaking up, and continued for about 50 million years.

We have much better records for temperatures and CO_2 levels, with much better dating, for the last 650,000 years. That's because we have ice core records going back that far. (Which implies, of course, that it was at least a bit too warm for ice in otherwise long-frozen regions before that.) Trapped air bubbles in the ice give us past CO_2 levels directly and temperatures indirectly, via the ratios of various isotopes.

These records can be correlated against current and fossilised tree rings, other fossils and, for recent centuries and decades, thermometer and satellite records to give us a quite good picture of the climate – with close enough analysis, even a general idea of the weather.

The last 650,000 years, during which our line within the once-diverse genus Homo evolved and developed to full humanity, have been marked by about half a dozen Ice Ages, in which ice advances about halfway over the Northern Hemisphere, less so over the Southern Hemisphere. Greenhouse gas levels drop to about 180ppm and sea levels drop by many metres. The Ice Ages never approach Snowball Earth levels – thanks, worms! – but must have been a huge challenge to our never-very-numerous direct ancestors.

Between the Ice Ages were warm, briefer interglacial periods like the current one, which began about 11,000 years ago. It seems to be no accident that agriculture and civilisation developed during this type of period, with the planet marked by very large temperate zones. (Temperate, of course, by human standards: warm enough for robust agriculture, too cool in most regions for many particularly dangerous diseases and pests, such as malaria-carrying mosquitoes.)

In interglacial periods greenhouse gas levels have, in the past, risen to about 280ppm – which is also the level found in recent times, just before industrialisation started sending them upward. Which implies that the next thing to happen without our intervention would have been a turn, a few thousand years from now, down toward another Ice Age. (Which, if we so chose at that point, we could have counteracted with controlled emissions of greenhouse gasses.)

During interglacial periods ice retreats to approximately current levels. Though we don't know exactly, it seems at least one such period was about 1°C warmer than today and marked by much less ice on Greenland and the West Antarctic ice shelf than we see now, with sea levels about 7 metres higher than today. While this would be a most uncomfortable, but perhaps survivable resting point for our current civilisation, warming we've already incurred plus warming guaranteed by past emissions doom us to warming even beyond that.

What does all this mean for life? For the better part of a million years, life has evolved in conditions ranging from colder Ice Age conditions, about 5°C below today's levels, to the moderate conditions that we'll soon leave behind as warming proceeds. Earth's current species are not that far from their cold-adapted predecessors, but not nearly as ready for temperature rises greater than about 1°C above recent pre-warming levels – meaning about 1°C higher than has been experienced for the better part of a million years. Significantly and quickly rising temperatures are likely to cause a great and profound die-off, with a long evolutionary path then required for speciation into the niches of a significantly warmer environment than has been seen for many hundreds of thousands of years.

(In extensive online and offline reading about climate change, I've never seen this concern expressed, let alone addressed. Either I don't get out nearly enough, or we need a step change in the focus we give to these problems and their implications. Or both.)

In the big picture, we are catching the Earth's climate near a recent (last million years or so) peak of CO_2 levels and temperatures and, virtually overnight, adding nearly 50% already to CO_2 levels while chopping down half the forests that, along with the seas, act to store excess CO_2. This guarantees more than 1°C of total warming already and threatens carbon sinks[1] in the environment that have the potential to send us back to Jurassic Park conditions – with a long period of extensive desertification and relative sterility, which I call Hot Earth, before life can diversify again to take full advantage of conditions. (No velociraptors again soon, sorry.)

Population growth

The current interglacial period opened the door to a rapid increase in human numbers around the world. At its beginning, stone tools gave humans a big advantage over competitors, both herbivore and carnivore. (Humans, omnivorous, compete with both.) American Indians, still in the Stone Age when Europeans "discovered" America, are a good example; stone mortars and pestles allowed them to process normally inedible acorns into meal; stone-tipped arrows and spears aided hunting. Humanity's growing dominance in this period is shown in our ability to domesticate wild food sources, such as

---

[1] "Carbon sink" is a broad term for natural features that cool the Earth, that absorb CO_2, or that have the potential to release CO_2 if damaged. For example, snow and ice cool the Earth by efficiently reflecting sunlight; new forest growth absorbs CO_2; forest fires release CO_2.

grains; to domesticate prey, such as cattle; and even to domesticate competing carnivores, such as wolves and wildcats.

Humans began to undergo what one can call the Petri dish phenomenon. A favourite parlour game of biologists is to fill a small, sterile glass dish with a more or less nutritious jelly, unleash some bacteria in it and watch the results. (Which are as follows: the bacterial population expands rapidly until the nutrients give out, then the bacteria all die.)

Biologists and students can do this enthusiastically over and over. Penicillin was discovered when some foreign bacteria blew into some Petri dishes where the biologist Alexander Fleming was playing exactly this game.

Petri dish bacteria also frequently foul their own nest, so to speak, with waste products. If they don't run out of food first, their little environment becomes too toxic to survive in, or at least to thrive in, causing various interesting dips and revivals in population as the local environment and the bacteria respond to each other.

Humans in a given region often behave identically to bacteria. We grow unrestrainedly in numbers, eat all the jelly – consume all of one or more crucial natural resources – and then we fight, flee and/or die, often crashing to a fraction of our former numbers.

In nature, where there are multiple species in an environmental niche, they are normally limited by simultaneous cooperation and competition with each other; "co-opetition" is a word that biologists should have invented before business gurus did. Individual species engage in a complicated negotiation with their biological as well as physical environments for survival. (An important insight which underlies the far more ambitious Gaia hypothesis, described in Chapter 3.)

Because humans are currently so dominant, with our growth in numbers largely unmoderated by competition from other species, humans regularly undergo Petri dish-type population explosions and crashes in various distinct environments, with various kinds of technology enabling the phenomenon on an ever greater scale. For instance, one of the first known human civilisations, at the junction of the Tigris and Euphrates rivers in modern-day Iraq, triggered desertification that still afflicts the region today. But it's only in the current period that our populations and capabilities have grown to the extent that we are risking a population crash on a global scale.

These tendencies are not some nasty characteristic specific to civilised humanity but are intrinsic to many kinds of life, enduring across evolutionary changes and specifics of place and time. They're even reflected in the behaviour of human economies, which cycle through boom and bust with depressing regularity. We can only hope to manage these tendencies, not eliminate them.[2]

The curve of human population growth over the last 11,000 years is one of exponential growth, just like the "growing" phase of those bacteria in a Petri dish. It took until about the year 1800 to reach 1 billion people; more than 100 years to add a second billion; about 50 years for the third billion, and we are currently at 6.5 billion people and adding nearly one billion people per decade.

There are actually three overlapping population curves reflecting what are, from an environmental point of view, three very different kinds of people: hunter-gatherers, whose maximum sustainable population on Earth may be in the low single-digit millions; agriculturalists, who could sustainably number perhaps a billion worldwide; and industrialised peoples. Though only about a billion people live a fully industrialised lifestyle, industrialisation affects agriculture worldwide, allowing the planet to support its current population of 6.5 billion people today. Whether this could be achieved was in doubt just a few decades ago, and we can still question whether it's sustainable even at current populations today.

Interestingly, hunter-gatherer lifestyles form the great bulk of humanity's past; agriculture is recent and not in all ways an advance, leading for instance to a poorer diet and stunted growth compared to either hunter-gatherers or industrialised peoples. And few hunter-gatherer peoples in history have voluntarily adopted an agricultural lifestyle; agriculture has been invented independently about a dozen times in world history, then carried throughout one or more regions by its numerous but perhaps discontented and undernourished practitioners. (The Basque people of northern Spain and southern France, famously irascible, are apparently the last relatively direct descendants of European hunter-gatherers otherwise displaced by farmers. Both their genes and their language hearken back to an otherwise lost heritage.)

The crises we will encounter in this century can be mapped to these modes of human existence on Earth, all operating simultaneously today:

- Hunter-gatherer man, armed with industrial age weapons – a modern fishing trawler is a powerful weapon indeed, to a fish – has overfished the world's oceans past the point of collapse for more and more fisheries;

---

[2] I owe much of the perspective informing the discussion here to Jared Diamond's excellent *Collapse* (2005), though the emphasis on fundamental and extensive similarities between humans and bacteria is my own.

- Agricultural man worsens the fisheries problem by polluting the seas with fertiliser runoff, creating expanding dead zones – and overuses the world's rainfall, snowmelt and aquifers to the point of collapse for the world's granaries and stockyards;
- Industrialised man worsens the fisheries problem further by acidifying the seas with CO_2; worsens the agricultural problem by wasting water and paving over productive land; and burdens the air with excess CO_2 as well, setting in motion a runaway chain reaction of warming that threatens to pull the rug out from under every element of our way of life at once.

At the same time, some tendencies of humanity give us grounds for hope in the face of the very problems so endemic to our nature. Educate people enough, and let them achieve a decent lifestyle, and they limit their own fecundity – to half of replacement levels in some countries today. It's this side effect of mass (relative) wealth, not war or mass starvation, that is projected to see the population peak at about 9.1 billion people in about 2050.

And we often do act intelligently as a group, curbing self-interest for the good of all. People have solved and cleaned up pollution problems before, from smog, to nuclear testing by-products, to CFCs that tear holes in the ozone. (Humanity narrowly escaped disaster with CFCs, which are also powerful greenhouse gases. The level of disbelief and unwillingness to act still found in current discussions around climate change shows that we have not learned as much as we should have from this.)

We would probably even solve climate change, if it operated gradually enough. The problem is that the global environment contains traps – carbon sinks – that have already begun to spring on us. Thus, runaway climate change.

Deforestation

One of the major and regrettable tendencies of environmentalists is to romanticise our less technologically skilled predecessors and current compatriots. A single example should help put paid to this proclivity: Genghis Khan, who killed millions, had set out to eliminate every city in China to open up pastureland for his horses until his accountant pointed out that letting the people live, and pay taxes, would be more profitable.

Past leaders did not, as a group, demonstrate sensitivity to nature nor brotherhood with their fellow humans to a noticeably greater extent than today's.

One of the little-noted accomplishments of our forebears, industriously exercising themselves over the millennia, was to raze about half the world's forests. (According to current measurements; further research could find that losses have been even greater.)

American Indians, sometimes portrayed as living in a kind of mystical balance with nature, burned forests to flush out game and to make room for acorn-bearing oaks, which they tended assiduously – and whose output they energetically processed into meal. Europe's monks happily spent the Middle Ages draining swamps – what we would preserve as "wetlands" today – and cutting down every tree they could get their axes on. What we today call "old growth forests" are evil places indeed in European folk tales!

Forests have long been – and, to a degree that we too often forget, still are – destroyed gradually for fuel. In fact, from a big picture point of view, the world's coal and oil reserves are just wonderfully compacted and useful extensions of the world's supply of firewood. We reached Peak Wood a while ago; the exploitation of coal and oil kept most people from feeling the effects, but shortages of firewood for fuel today afflict a large number of people.

Greenery today absorbs about one-fourth of greenhouse gas emissions. (The seas absorb another one-fourth.) Unconstrained, greenery grows in intensity per acre and extent as CO_2 rises, reducing the net increase in CO_2. But damaged or destroyed forests lose their ability to respond to the opportunity for growth afforded by greater CO_2 concentrations.

Deforestation also contributes to water shortages and sets the stage to worsen the effects of less reliable rainfall per season and "bursty" rainfall caused by climate change. Where a forest traps a winter's rainfall, then releases it gradually during the spring, non-forested areas let water run straight through them, then dry out quickly as soon as spring begins.

Deforested areas also contribute to diminished rainfall, thereby causing drought and desertification. Forests steadily lose water to evaporation, which adds vapour to passing clouds, triggering rain. Deforested areas allow clouds to pass by until they reach a forested area, mountains or the sea.

Today, we're directly losing about one percent of the pre-warming extent of the world's forests every year by direct human action – forests being cleared for cropland (including to grow biofuel crops), for grazing for cattle, for construction and for roads.

At current rates of destruction, half or more of the world's current forests – an additional quarter of the one-time total – will be lost in the next 25 years. This will worsen climate change and further reduce the potential for recovery.

In later chapters I'll describe how forests are further damaged by climate change and are beginning to tip toward seeming self-destruction, from the twin, related threats of parasite damage and fire, contributing to runaway climate change. What's worth keeping in mind throughout is that the world's greenery, and the world itself, have been "teed up" for this mortal threat to great swathes of humanity by ongoing deforestation of half the world's forest extent caused by people over millennia.

Lack of fresh water

Another poorly understood element of climate change is the extent to which the groundwork for drought and desertification – which will greatly reduce agricultural potential throughout the rest of the century, bringing forward Peak Food – has been laid by overuse of water supplies and increasing pressure on them from humanity's current population explosion.[3]

As mentioned, humans have a tendency, going back to bacterial days, to overuse resources. To agricultural man – which we virtually all are, directly or indirectly, including the one-sixth of people who are also industrialised – the most important resource around is fresh water. A human being only needs one or two dozen litres of water per day for drinking and washing, but about 1000 litres of water are needed to grow a kilogram of grain. So by far our greatest consumption of water is by, in effect, eating it.

Humans tend to first populate well-watered lands, then to overpopulate them and push out to marginal lands on the fringes of the prime areas. Use of the marginal lands often occurs during episodic wetter periods. People then overuse and ruin the marginal lands, starting desertification; and then must retreat as advancing deserts and depleted soils ruin first the marginal areas, then the once-rich heartland. This cycle, being seen in China right now – with Beijing itself at risk – is simply the Petri dish all over again.

When something is cheap or free, people tend to find ever more creative ways to overuse it. Grain needs immense amounts of water, but livestock – especially cattle – need large amounts of feed, representing a great deal more water, plus water for cleaning pens, meat processing and even packaging. Americans, the biggest beefeaters, "eat" far more water per person than other peoples, but

---

[3] I'm indebted to the brave and excellent book, *Plan B* by Lester Brown (several editions, beginning in 2003), for first presenting this complex connection to me, and many others, in a coherent fashion.

the developing world is closing the gap. With grain markets increasingly global, the resulting impact on grain prices is a worldwide issue.

Rainfall is one source of water, but only some of it falls in the spring and summer, when it's most needed to help crops grow. It's best for crops if the rain falls as snow instead and accumulates as snow or ice. It then pours forth steadily through the spring and summer in the form of snowmelt or glacier melt. Farmers are thus steadily and reliably delivered an entire winter's snowfall, which is far more predictable than each week's rains; and, while dry years are still a problem, they're at least known about before planting season, making some adjustments in crop choices and overall investment possible.

Generally reliable snowmelt is found, for instance, in California's main Sierra Nevada mountain range. ("Sierra" means mountain range and "Nevada" means snowy.) Snowmelt provides two thirds of the state's water supply. Each spring and summer, snowmelt feeds the rich fields of California's Central Valley, a huge breadbasket for the whole of North America and beyond. California produces approximately 20 percent of the agricultural output in the entire United States. But the annual Sierra Nevada snowpack seems to be in steady decline; at this writing, California is in drought, which may become its usual status for a very long time to come.

Dams can potentially hold several seasons' water. When dams are good, they're very good; water releases can be scheduled and every drop (that doesn't evaporate) can be put to use. On downstream farms, long-term investments stretching over decades become possible. But when dams are bad, they're very bad; multiple years of less than anticipated rainfall can abruptly cut once-reliable deliveries to nothing, laying all those investments to waste. This rarely occurred before climate change, but is now a threat in many places.

Populations increase on the steady supply of food grown from dam water, as well as water for household and industrial use. Hoover Dam in Nevada, steadily supplying much of the American Southwest with what would otherwise be highly seasonal Colorado River water, is a spectacular example, sustaining millions of acres of farms and millions of new households throughout the Southwest. But today Hoover Dam's reservoir, Lake Mead, has fallen to half of accustomed levels.

When dam water gives out, the populations dependent on it are left, excuse the expression, high and dry. Before efficient worldwide transport, people would have starved, then migrated or died; today the worldwide market in grain buffers the impact, though at the cost of worldwide increases in food prices. Eventually, though, collapsed food sources act as a localised constraint on population sizes, with the impact decreasing in concentric circles outward from the collapse.

Glaciers act much like snowmelt and dams, but with centuries' worth of water storage rather than one year's or a decade's worth. Though not as controllable as dams, they're more reliable due to their long-term nature – and dams are often built downstream from glaciers to give the best of both worlds. China and India's billions of people are largely supported by Himalayan glaciers which are now in long-term decline; Peruvian crops and millions of people in Lima and other cities are in danger from the ongoing decline of Andean glaciers.

Even better than glaciers, though, is groundwater. Aquifers – natural underground water storage – are highly efficient, capable of being tapped for water as when needed and incurring little evaporation loss. Any sensible region or nation would limit withdrawals from aquifers to the rate at which they recharge from rain and snowfall, creating a truly sustainable situation, but almost no one does this. (Petri dish, anyone?) People just dig deeper and deeper wells, assisted, since World War II, by ever-cheaper electric motors. More and more wells worldwide are now becoming uneconomical to operate, delivering silty, salty or polluted water, or just giving out completely.

Cities are vulnerable to regional food shortages and to direct effects from aquifer exhaustion. Many draw groundwater for municipal use; not only can this source give out, but coastal cities – the overwhelming majority – find that salt water seeps in to replace the diminished fresh water, ruining the entire supply. Even the small sea level rises and increase in intense storms caused by climate change to date can turn this from a long-term threat to tomorrow's headlines.

If the climate and populations were roughly stable, the world would still have a very serious water problem as aquifers collapse throughout the coming decades. Fast-rising populations are enough to make it a crisis. The additional impact of climate change – especially non-linear, runaway climate change – threatens to combine with this pre-existing water crisis and the continuing population explosion to create a worldwide disaster of very high global food costs, regional starvation and forced emigration.

Increasing vulnerability to rising seas

One of the ironies of climate change is that the negative effects are all deferred; a load of CO_2 dumped in the atmosphere today takes decades to fully affect global temperatures, and further decades, even centuries, to fully affect the cryosphere (Earth's ice and snow) and thus sea levels.

Yet when discussing climate change, the only impact that's truly compelling right across environmentalists, business, government and the citizenry at large is the last one to occur: sea level rises. You can talk about the record-breaking summer heat wave of 2003 killing more than 10,000 Frenchmen until you're blue in the face; or recount the difficulties for plants and wildlife in shifting habitats by many metres a year and get resigned shrugs and people checking their watches in response. But start describing recent estimates of sea level rises averaging a mere centimetre a year – 4 inches a decade, a metre a century – and you'll get their full attention. Add the growing likelihood of a metre's sea level rise by 2050, recently mentioned by several well-informed scientists, and something resembling panic ensues.

Why is this? Partly because changes in sea level, unlike changes in temperature, are not a matter of degree (sorry) – hotter vs. cooler, masked by daily and seasonal variations – but of kind: habitable or not, underwater or dry. And also because coastlines are by far the most densely populated, and the most vulnerable, topography on Earth. In the US – even more coastally populated than many other countries – 70 percent of the population lives within 50 miles of the shore.

Sea level rises also compel attention because their impact is amplified by a largely quiet and unexcitable group who are yet some of the most dangerous people on Earth, whose work will bring disastrous impacts to millions and tens of millions of coastal residents and investors in the coming years: insurance underwriters. Coastal property is the most valuable in the world, worth literally trillions of dollars; yet more and more of these properties will see their value crash to near-zero as the risks that are present today, and increasing tomorrow, are fully reflected in insurance premiums, with insurance very suddenly becoming prohibitively expensive or just unavailable in once-prime areas.

We'll talk about the hotly debated potential for sea level rises, and touch on their impacts, in later chapters. But why are coastlines so populated? And why are they particularly vulnerable?

Much of the answer has to do with water, salt and fresh both. Ports are great for trade, and river mouths have both fresh water and fertile land, annually renewed by new silt deposits, making them prime agricultural land.

Riverside cities can be founded upstream as well, but port cities have extra stimulus for growth. Downstream cities also have the easiest access to aquifer water, with the water table close to the surface. Only sites near sea level – the great majority near the sea itself – offer the shortest distance to the top of the water table.

All these factors applied before industrialisation, determining the initial location and growth of cities. Industrialisation – even in countries yet to fully industrialise – largely determined which cities became behemoths. With industrialisation, transport of food and water to supply cities became easier. (The damming of the beautifully scenic Hetch Hetchy valley in Yosemite National Forest, to provide water for the city of San Francisco hundreds of miles away, in the 1920s is just one example.)

With industrialisation, many fewer people are tied to the land; "how you gonna keep them down on the farm?" becomes a rhetorical question. In post-Revolutionary America, for instance, 90% of people lived on farms, which were spread out over the arable land. Now that fewer than 10% live on farms, people live where they want to live and where jobs are more likely to be – which is increasingly in cities, and most often in cities near the coast.

The move from rural areas to cities – urbanisation – is one of the changes that marks industrialising countries. As most countries in the world move into a pipeline toward industrialisation, these changes are becoming global. It was recently announced that 2009 is the first year that the world's urban population exceeded its rural population – which also means more and more movement to the coasts.

This mass proclivity puts people in places that are particularly vulnerable to climate change for three reasons, one obvious and two not so obvious:

1. **Submersion**. The obvious one – once the high tide line reaches your front door, you can't live there any more. What few people realise, though, is that, in many areas, a sea level rise of just one foot may move the coastline in by 50 feet or more.
2. **Loss of fresh water**. Many cities support their population's needs from nearby aquifers that are easily spoiled by seepage: the seawater pushes in as it rises and is pulled in by the aquifer's gradual depletion.
3. **Storm damage**. This is the agonizing one. People build right up to (or even beyond) the line at which their property could endure a 100-year storm or flood. As climate change brings "100-year" storms every decade or two, the boost given by even small sea level rises accelerates the storms' impact, necessitating expensive protections for, or abandonment of areas that are otherwise safe.

Not only people, but agriculture and industry are clustered near the coast. Los Angeles, for instance, a low-density city of 4 million people

(10 million in Los Angeles County), is surprisingly industrial – and surprisingly agricultural. Loss of land for people also means, in nearly the same proportion, loss of industrial capacity – and a lesser, but still significant loss of agriculture, worsened by saltwater intrusion into aquifers well ahead of the actual mean high tide level.

There are also innumerable pollution disasters waiting to happen. About half of America's oil refining capacity is on the Gulf Coast, just waiting to be flooded – pouring all sorts of pollution into the already strained Gulf of Mexico.

So the disruption caused by rising seas is agonizing in more ways than one. The core is the gradual loss of land to sea. But ahead of that definitive loss is a much more extensive front of loss of access to freshwater and increasing vulnerability to storm damage. And surrounding all of that is a dense fog of uncertainty about how much sea level rises and storm damage will increase in a given period of years or decades ahead, which affects people most immediately through the sudden loss of insurability for existing and proposed development.

New Orleans is a good and familiar example. Once it was damaged, much of it didn't make sense to rebuild – but decisions are being made by benign neglect, not in a sensible or fair way. And even the part that was rebuilt may be too expensive to defend from, or rebuild after, some potential further disaster that may be just a decade or two down the road.

The core issue is the still-increasing degree to which our human, industrial and agricultural capital lies near coastlines and is therefore so very vulnerable to the rising seas, loss of aquifers, storm damage and loss of insurability caused by climate change.

"Western" lifestyles

The most important megatrend for climate change is the move to "Western" lifestyles – so called because industrialisation first arrived in Western Europe and the US, as well as such British Commonwealth countries as Canada, Australia and New Zealand, followed by Japan, Korea, Taiwan and Hong Kong.

A "Western" lifestyle is marked by high incomes, good health and long life spans. It includes widespread use of cars and jet planes and high consumption of meat, especially beef. A tiny minority of people farm, with the great majority working in manufacturing or services.

From a really big picture point of view, the Western lifestyle combines the relative security and higher population densities of an agrarian lifestyle with some of the independence and freedom of a hunter-gatherer lifestyle. (Closing a business deal is sometimes called "going in for the kill", just for one example.) For all of these reasons, the Western lifestyle is compellingly attractive to those who don't yet have it.

Russia took to Communism to close the gap with the West; China adopted the almost unbelievably wrenching (and still continuing) one child per family policy in an attempt, so far successful, to vault a billion people into Western lifestyles in just a few generations.

In marketing terms, the Western lifestyle can be regarded as a product, and an extremely successful one. In about 250 years since industrialisation began it's penetrated one sixth of the Earth's fast-growing population. One-sixth is the fraction usually considered the "tipping point", beyond which a phenomenon is likely to become ubiquitous. Proving its attractive power, the Western lifestyle is about to add, in the populations of China and India alone, another third of Earth's population to its "market share".

This, and other comments in this book, are not meant to "blame" China or India. The rapid emergence of a country as large as the rest of the developed world into that world was always bound to have big effects; the fact that it's occurring just as climate change becomes the world's #1 problem multiplies the impact.

The Western lifestyle seems addictive; even European countries proud to have reduced their carbon footprints seem, on further analysis, to have largely just shifted their emission-producing activities to other countries. There's an informative online presentation, The Story of Stuff, that outlines how Westerners are both practitioners and victims of planned obsolescence and other consumption-increasing practices.

Industrialisation and the Western lifestyle are, unfortunately, almost wholly responsible for greenhouse gas emissions and therefore for climate change. They also enable and accelerate the accompanying megatrends described here. The greater populations made possible by industrialisation emit CO_2 by deforesting (even) faster than agriculturalists could manage, and by emitting CO_2 directly from factory smokestacks, car tailpipes and jet engine exhausts. (As well as CO_2 and methane from the guts of farm animals, especially cattle.)

Unfortunately, people and governments tend to indulge in harsh cost-benefit calculations and wrenching political debates before deciding to pollutants. Smog in Western cities became quite severe before being brought under control, most notably in the 1970s. Much of Asia today suffers from even worse smog, coalescing into giant Atmospheric Brown Clouds (ABCs), that could be easily reduced by technology and practices known and used in the West for two generations; but the needed regulations, laws and enforcement are deferred in the name of progress.

The trick in combating climate change is to offer people the Western lifestyle that nearly everyone wants to keep or attain without the emissions and deforestation that accompany it. Unfortunately, unlike smog, there's no easy and relatively affordable set of fixes waiting on the shelf to be adopted against climate chage. Quite the opposite; as global oil production plateaus and prices seem set to rise, energy generation shifts to far more damaging coal and heavier, hard to extract oils such as the aptly named tar sands.

All this is coming at a crisis point for the Western economic model, with the world economy only limping along due to the continuing growth of, a fact beyond irony, Communist Party-led China.

Any attempt to battle climate change will have to begin by placing the blame for it squarely where it belongs – on Western lifestyles – and either confront their addictive power directly or, more likely, reframe it by using industrialisation's economic output to build a new, non-polluting infrastructure.

Extinctions and declines in numbers

A large degree of destruction of the natural environment – even a wipe-out of more than half of all known species – is a grave danger of runaway climate change. In fact, it seems almost unavoidable.

Much of this loss is teed up by pre-existing human impact on the natural world. As top predators, people have always had a big impact. The world is full of examples of plants and animals wiped off islands, continents or the entire planet by their interactions with humanity. (Lions in Britain, the horse in the Americas – until re-introduced by Europeans after Columbus – and the dodo bird being famous examples of each in turn.)

But the encroachments of industrialisation and growing human numbers have accelerated humanity's impact on the natural world, with conservation efforts now focusing on the preservation of relatively small areas of habitat for endangered species.

In this focus on extinction, as important as it is, we somewhat miss the bigger picture. The number of creatures per species matters almost as much as the numeric count of species. The land and seas once teemed with life that has lost out to human expansion. Extinction is just the last tragic marker on a long and sad path from biological richness to devastation of ecosystem after ecosystem.

Damage to the natural world is a culmination of other megatrends discussed, in particular population growth and deforestation. Both these trends helped create the need for nature preserves – and both are pressurising the preserves that do exist, both official and unofficial.

From ground level, the remaining wild areas still seem vast and untameable. But, with humanity controlling all the key variables, they are really just large theme parks. (As far as true wilderness is concerned, the entire planet is really now just an extra-large theme park.) As with climate change itself, we continue to treat as a series of local difficulties what is really a global and long-term problem. (Though the non-profits and their supporters have caught on: it's no accident that where we once had America's National Audubon Society or Sierra Club as leading conservation organisations, we more often now see the World Wildlife Federation.)

And, even with climate change just kicking into gear, we already see global and even existential threats. Frogs and toads worldwide are in serious, rapid and unexplained decline. Large numbers of honeybees are dying year by year, reducing pollination and thereby threatening important parts of the world's food supply.

The seas give examples of the losses we are suffering. Because the seas are so vast, complete extinctions are rare. Yet recent research into old records is showing how the richness and diversity of first freshwater, then ocean life has been steadily reduced by overfishing, to the point where the seas are now, biologically, a shadow of their former selves. Yet if they were allowed to regenerate and protected, they would be far more productive for humans than they are today, and with less effort and expense.

These declines demonstrate something that should have grabbed humanity's attention much more forcefully many years ago – which also would have led us to slow down or prevent climate change: species are interdependent, including our own. As we pressure the most vulnerable species ever more harshly, we are quite likely to find environmental webs unravelling quickly, with the lost diversity not to be replaced without significant new evolution requiring many millions of years.

Climate change, as it transforms each of the remaining isolated and disconnected habitats in turn, threatens to finish the job for thousands of species, unable to migrate with enough speed and in sufficient numbers to potentially newly suitable habitat that is itself in flux. Species already at high altitudes or high latitudes are beginning to be blown right off the planet. The process is no less violent in its effects for playing out over decades rather than in a sudden Götterdämmerung.

This book gives little focus to problems in the natural world, but only because we have so much ignored the problems in nature that we are now directly threatening ourselves. If we can't find the motivation to save ourselves, Bambi is hardly going to get a fair shake.

Part of the threat of climate change is the threat of leaving future generations a biologically bereft planet. Any solution for people must preserve what remains of, and significantly restore, the natural world as well.

Megatrends and climate change

Given these challenging underlying megatrends, there has, until just the last few years, been impressive progress on a number of global problems tied to them.

The Green Revolution in crop growth roughly doubled agricultural productivity, lifting the spectre of starvation for hundreds of millions of people and allowing the world's population to grow by several additional billions. The number of people in extreme poverty – defined as living on less that $1 a day – was one-third of humanity in the early 1990s, but is now only one-fifth. (This is still more than a billion people – the "bottom billion" – but the progress is remarkable.) This share was recently projected to drop further, to one-sixth, by 2015. But that was before the global credit crunch and recession hit, and did not take into account recent news – and new views, such as those expressed in this book – on climate change.

The number of people living without access to clean water, also more than a billion, has been gradually reduced in recent decades. As has illiteracy, at about 750 million people today. The Millennium Development Goals, still being pursued today, include progress on these and several other markers of poverty and ill-health.

Unfortunately, some of that progress is now being reversed. The number of undernourished people in the world's least developed countries is estimated to have grown in the last ten years, from about 800 million to roughly a billion – and is projected to grow by the same amount in the next ten years.

The question of immediate vs. underlying causes can be entertainingly debated with relation to almost any problem or crisis. However, given the megatrends described here, this century would already be set to become a difficult and even potentially disastrous one for much of humanity, even if all the excess greenhouse gases in the air and acidity in the seas were to be neutralised overnight. The problems of increasing population, deforestation, overuse of fresh water and damage to the natural world are that bad.

The other megatrends described here – the fact that we're near the peak of an interglacial warming cycle and our tendency to live near coasts — predispose us to problems from climate change-caused warming and sea level rises.

Given that we have so badly neglected these interrelated problems, and that global warming and oceanic acidity are bearing down on us like a freight train, we have teed ourselves up for an intensifying series of local disasters that are already, in fact, beginning to occur.

A particularly striking example is the brushfires of early 2009 near Melbourne, Australia. This region has always been a marginal environment for people and has always been prone to brushfires. (The Aborigines used to start fires deliberately in the area as a preventative measure. "Let's give it back" indeed.)

So the recent fires there were not a surprise — but their ferocity, coming after years of record drought, then days of blistering temperatures reaching a record 46°C (115°F), was. Fires swept from one horizon to the other with shocking rapidity, contributing to a tragedy which saw nearly two hundred people die.

The more gradual effects of drought are also marked in the region. Towns are disappearing as local inhabitants pull up stakes and leave. In a particularly heart-wrenching trend, increasing numbers of farmers and stockmen, running out of hope, take their own lives.

This area of Australia is a harbinger of problems to come in other parts of the world. The combination of the megatrends described here and the increasing impact of climate change will set off many such accidents that are increasingly just waiting to happen. As they actually occur with greater and greater frequency, we will be left desperately looking for answers.

Effective solutions will be complex because they will require addressing longer-term, underlying problems and basic human tendencies we have allowed ourselves to ignore. But those of us who want these broader concerns to be taken into account may be asking too much if we try to raise them at points where urgent action is required.

But you, gentle reader, have the gift of a bit of time to think. Keep these longer-term concerns in mind as we discuss the crises of climate change and runaway climate change that are now emerging — seemingly suddenly, but actually with deeper roots, as you will have seen in this chapter, than is generally recognised.

Chapter 2. Climate Change 101

> In this chapter
> - The subtlety of climate change impacts to date
> - "Commonsense" evidence for climate change
> - "Scientific" evidence for climate change
> - Causes of climate change
> - The persistence of emissions
> - The Kyoto Protocol and its effects

Climate change is a particularly abstract and subtle challenge to humanity. After all, the day-to-day weather changes all the time, and even the longer-term climate shifts on various cycles. How can we know that something that should concern us is indeed happening – and understand it better, as a guide to responding effectively?

In this chapter I set out the basic argument that climate change is, when assessed by people's individual senses and life experience, quite subtle – compared to, say, air pollution, which forces itself on one's attention quite dramatically. The effects of climate change, in the early stages that we've experienced up to now, are simply an intensification or reduction in things that happen all the time without climate change, and are therefore quite hard to perceive as symptoms of a truly serious problem.

So here I divide the evidence of climate change into "commonsense" and "scientific" components. The first part is to root the core claim, that the Earth is warming, in evidence anyone can understand and hopefully come to agreement on. This should build confidence in, and patience with, the more "scientific" evidence, which the rest of us have to take on faith as it's proclaimed by scientists – who, unavoidably but unhelpfully, make mistakes, argue and change their minds.

I further divide the "scientific" part into fact and theory. Scientific facts can become agreed and settled, but scientific theories are just that. We can accomplish great things using them – as we see every time we ride in a car or make a mobile phone call – but we always have to acknowledge that they can be replaced completely; or, more commonly, superseded by a broader understanding that reframes, without completely changing, previous perspectives. Keeping facts and theories separate helps us understand where to demand agreement from others and where to keep an open mind, even while – as we must – using theories to guide action.

The overall shift we refer to as part of climate change is a matter of degree – a bit over a single degree Fahrenheit (0.8°C), so far – not of kind, yet.

The Fahrenheit temperature scale was constructed to encompass the temperature range most people – well, most people near Daniel Gabriel Fahrenheit's Dutch home, anyway – live in most of the time: from somewhat below freezing (0°F, -17.5°C) to a very hot day indeed (100°F, 38°C). The temperature change experienced so far due to climate change is only a bit more than 1% of this range. One can barely feel, if at all, a single degree of change in the temperature. It's far less than the temperature difference between day and night, or between mid-summer and mid-winter. And it's less than regional shifts in climate that have always occurred, for a variety of reasons that don't imperil humanity – though they can certainly damage a local economy.

The small, but global shift caused by climate change has also occurred over a long time scale, a period of a century. (The current warming started in about 1910.) So climate change, while critically important, is so far driven by a shift too small for a person to easily notice, spread out over a timeframe longer than a typical human life.

Also, climate change tends less to cause extremes of heat than to level them out. It makes nights warmer more than it does days. It warms the cold, nearly uninhabited polar regions more than the already hot, but more densely populated, tropics. So it lessens the difference between extremes, which is harder to notice, rather than increasing them, which would be much easier to notice. It's only when climate change is well along, as it is now, that it begins causing more and more noticeable instances of extremes as well.

The places where climate change has made the most visible difference, and is now causing rapid change, is in formerly ice-covered cold regions. Despite the fact that few people live in these regions to experience the change directly, they do provide some tangible evidence, in the form of retreating ice, to help convince people the problem is real.

Whether or not climate change is runaway is a similarly subtle point. The very definition of "runaway climate change" can be argued, and the determination as to whether a given definition is met is based largely on still-uncertain evidence.

This book will demonstrate that climate change is real, provide a strict definition of runaway climate change, and show that runaway climate change – even by this strict definition – has already begun.

Why can't we all just get along?

The subtlety of the changes involved is perhaps the major reason climate change is so hard to convince people of. Research shows again and again that we live mostly in our bodies and respond strongly to what our senses tell us directly and our emotions impel us to do – not as much in our brains, responding to reason.

A good example of this is the debate over the damage to human health caused by cigarettes – damage that, as with climate change, is long deferred from the acts which cause it. It's only now, after many decades of argument, widely agreed that smoking causes lung cancer – the arguments back and forth are similar to what we are now undergoing with climate change. And discouragingly, even with the conclusions on tobacco generally agreed – smoking kills – about a quarter of the population worldwide still smokes.

Those who believe that climate change is real, and a real threat, often believe that those who disagree or demur are stupid. The truth is, most people just aren't built that way. We tend not to sacrifice physical comforts and pleasures – which are easily disparaged, but relate to our daily physical survival – against subtle, delayed, intellectually argued dangers.

It's only when the intellectual argument is translated into tangible evidence or social pressure – as has happened with cigarette smoking – that change happens. Change which usually takes a long time to come about, and which has an effect that is far from total.

The climate change believers fall prey to this themselves. Any purely logical approach to the dangers of climate change would first investigate the possibility that things are really bad – in this case, that climate change has already run away, tipping carbon sinks in nature, as described in this book. Yet the all too comfortable consensus du jour is that we can save ourselves while veering only slightly away from business as usual.

Some climate science advocates, as some of us describe ourselves, even argue, to the point of seeming ridiculous, that arresting climate change will be inexpensive, and imply or even state that it will be easy. Huge investments will be needed; *net* costs may be low, as green energy may well be cheaper, but tremendous cash, effort and brainpower will be needed. Clever re-allocation of costs can make the process as fair as possible, but it's yet to be proven that this will actually happen.

But it's even worse than that. Not only are people slow to believe things they don't sense, they – we – are even slower than that to actually do something. Hitler wrote out his evil plans in *Mein Kampf* years before launching World War II. Only a few, fortunately including Winston Churchill, took him seriously.

There are always a thousand things to worry about – for individuals, families, companies and governments. Only those that at least begin to cause real damage are likely to get attention.

There's even evidence that poll results on the seriousness of climate change as an issue depend on just how warm the weather is at the time when the respondent is asked! Those who already hold a strong opinion on either side tend to stay with it, but "swing voters" are influenced by what they are seeing and feeling in the moment.

So those of us who are, appropriately, deeply worried must be tolerant of others who seem slow to "get it". It's just as "normal" to be dubious and slow to act on climate change than to be easily convinced and quick to act.

Climate change is as great a challenge as we humans are likely to face, largely because of its subtlety. Despite the extreme urgency, convincing people of the extent of the problem, then convincing them to act effectively, will be a long, slow, frustrating process.

"Commonsense" evidence

With climate change being such a subtle phenomenon, and with the stakes so high, it's important to use "commonsense" evidence to make the case that it's actually occurring. "Commonsense" evidence is information that's available to ordinary people, that's understandable without special training, and that's thereby resistant to misuse or misinterpretation.

The best commonsense evidence of climate change is the worldwide retreat of glaciers around the world, especially glaciers in temperate or even tropical areas that are regularly observed by many people. As a group, these glaciers have been in retreat for almost a century, with their decline especially noticeable in the past few decades.

Though it takes place on a timescale of decades, the retreat of these glaciers can be verified by comparison to old photographs owned by a range of different individuals and in magazines, on postcards and so on – evidence that's almost impossible to fake.

These retreating glaciers are by themselves decisive evidence of global warming. There is nothing that could make all these glaciers retreat except worldwide warming. The glaciers alone make it clear that we have been experiencing warming for about a century, to a degree unprecedented for at least the last several hundred years. This is the only evidence one needs to prove significant and ongoing global warming.

If anyone wants to assert that global warming is not occurring – as a long-term trend lasting a century so far, and continuing today – they have to explain the long-term and continuing, even accelerating, melting of glaciers. They won't be able to. And if someone wants to assert that warming has stopped – not just that we're having a slightly cooler year or two, but that the trend has truly halted or even reversed – they'll have to show that the glaciers are growing again. They can't. (And my assertion, argued throughout this book, is that they'll never be able to; without unprecedented human action, every glacier on Earth is doomed.)

There is much other commonsense evidence as well. Growing seasons observed by farmers have changed. Planting times are earlier and tree leaves fall later – shifting by a bit more than one day per decade in the last several decades, more than a week in total, which is a lot in seasons that average only 13 weeks in length. But people who don't want to know tend to dismiss such evidence. They can't dismiss the retreat of ice.

Hunters and gardeners both have observed that habitats of plants and animals are shifting away from the tropics and toward the poles, or away from lower elevations and toward higher ones, as if living things were trying to escape heat. Pests that can only live or thrive with high temperatures, or with a lack of freezes, are extending their ranges into warming areas. (This is in line with climate change's tendency to level out extremes; the occasional hard, insect-killing freeze that once kept pests under control in temperate zones is among the first things to go with warming.)

The changes are most dramatic where it had been coldest – and warming is highest. In the Arctic region the number of days on which the ground stays frozen, supporting driving, has declined steadily for decades. Buildings, pipeline supports and other structures are twisting crazily in some areas as once-reliable permafrost melts.

Ordinary citizens have also long been involved in measuring temperatures. These thermometer records can be criticised as subject to error, patchy and affected by changes such as urbanisation, which causes localised warming. But careful analysis shows that the trend is universally upward – to an extent that's impossible to explain except by overall warming.

All of this evidence is enough to prove to unbiased observers that warming is indeed going on, to a scale not observed for at least many centuries. However, written records seem to indicate a period of warmer weather about 1000 years ago, referred to as the Medieval

Warm Period – though records for this past period are quite sketchy. Crops grew in new areas – in particular wine grapes, which are well known to be quite demanding as to temperatures.

The amount of change occurring so far in recent times is small, so commonsense evidence has not been enough, by itself, to prove that the current warming is greater than what may have occurred in the periods before thermometer records began, more than 200 years ago – and in the Medieval Warm Period in particular. It's taken careful study of other evidence to show that the current warming is indeed unprecedented in at least 10,000 years, and quite likely in more than a million.

But the most recent commonsense evidence should put an end to doubts. Scientists have claimed for decades that the North Pole's ice is thinning. Now we see the North Pole's ice cap getting significantly smaller. In 2008, for the first time ever recorded, both the Northern and Northwest Passages – over Russia and Canada respectively – were open. The North Pole's ice cap is expected to disappear in summer within a few decades.

This is a major geophysical change in our planet, caused by us. (There is no evidence that sunspots, orbital variations or volcanic eruptions, to name a few favourite non-human-caused alternatives, could be causing this.) It is evidence as unequivocal as one could ask for.

The string of record and near-record high temperatures in the 1990s and this decade had removed most doubt that we're in a period of record warming – and the changes at the North Pole should remove the rest. Sceptics have argued every measurement and gleefully celebrated any adjustment that seemed to move a recent high just out of the "record" category. But the sceptics sound increasingly desperate as evidence against their position accumulates.

"Scientific" evidence

Commonsense evidence is enough to show us that warming is occurring, but not the amount or cause. For that, we need to turn to science.

Science has access to places, instruments and analytic techniques that ordinary people don't. A rather spectacular example can be found right at the core – no pun intended – of climate change science.

Much of the long-term evidence for past climates comes from ice core data. In multi-million dollar projects, scientists have dug deep and used powerful drills to extract very long tubes of ice, containing deposits of ice – and trapped air bubbles – going back 650,000 years.

Analyzing these cores, scientists have learned a great deal about past climate.

Scientific evidence is both more powerful than commonsense evidence and in some ways less trustworthy. That's because the access, instruments and analysis, being specialised, are also subject to fraud or error.

Climate change science offers examples of this as well. Climate change first came to public attention in the 1980s, and sceptics attacked some of the strongest evidence: decades of thermometer data showing slight but steady warming. A key criticism was that the thermometer data, which tended to come from settled areas, wasn't truly global. So the results from a satellite that circled the Earth throughout the 1980s, measuring temperatures worldwide, were eagerly awaited.

And the sceptics seemed vindicated. The satellite data for the decade showed five years of slight warming – followed by five years of slight cooling. Yet Earth-based thermometers showed ten years of continued gradual warming over the same period. This seemed to disprove not just recent evidence of warming, but to put in doubt all the previous decades of thermometer data as well.

Eventually, an error was found – as the weather satellite circled the Earth, atmospheric drag slowed it slightly. It gradually spiralled closer to Earth, and its results were therefore time-shifted as well. When appropriate corrections were applied, the satellite data matched the temperature data very well.

But the sceptics had had their headlines. And even a sympathetic observer could hardly avoid doubts. If the lead scientists had been too dim to predict the atmospheric drag, which is just basic physics – and had to have it pointed out by others – how could anyone be convinced they had calibrated their instruments correctly, and on and on? The underlying trust needed to make the results of such a complex experiment convincing was badly damaged.

Ironically, the corrected satellite results were extremely valuable. They could be correlated with terrestrial thermometer readings, ice core analysis, tree ring records and other information, stiffening the entire body of data.

With all this and more now at hand, scientists are able to provide the specifics we need. Over the past 650,000 years – as far back as we have ice core data – global average temperatures seem to have stayed in a range of about 5°C, swinging between the depths of several Ice Ages and warmer periods. For the past 11,000 years, since the end of the last Ice Age, we've been in one of these warmer periods, near the high end of the range.

Global temperatures started rising above this level around 1910. In the ensuing decades, they rose slightly, then plateaued from midcentury. They then started to march upward steadily in the 1970s at about 0.15°C per decade – more recently at 0.2°C per decade. This has culminated in recent years that have seen record and near-record temperatures, threatening to take us beyond the range we've stayed within for most of a million years.

The most recent evidence – the increasingly evident collapse of the North Pole's ice cap, with record shrinkage in the last two years – proves that not only are record temperatures beginning to occur, but that temperatures have been steadily higher for long enough to cause landmark changes in nature.

But why is climate change occurring? For that, we need to go beyond scientific facts to scientific theory.

The causes of climate change

Modern science has undermined the idea of "proving" anything. Statistically, we can establish that two things are correlated – they rise and fall together – but not, without great difficulty and disagreement, causation: that one thing makes the other happen.

To show causation is not only difficult; there are some who will claim, from a philosophical perspective, that it's impossible.

But science, while recently discounting causation, is ironically the greatest guide to causes and effects ever found. It's science that enables us to create most of the wonders – and all too many of the horrors – that distinguish our modern age.

For climate change, we have a strong case that it's caused – there's that word – by greenhouse gases.

Over a period of a few centuries, a succession of scientists contributed to our understanding of climate. First they established that the only significant source of heat on Earth is the sun; internal heating only contributes a few degrees above absolutely zero. Then they worked out that much sunlight that reaches our planet reflects right back from the Earth, especially from clouds, ice or snow – almost as effectively as if it had been blocked from reaching Earth at all. The rest of the sunlight is absorbed as heat, much of which then radiates back out into space, only a bit more gradually.

What keeps the surface of the Earth warm are greenhouse gases – water vapour (hydrogen dioxide, or H_2O), carbon dioxide (CO_2), methane (CH_4) and, to a lesser extent, others.

These molecules have loose atomic bonds – H-O bonds, C-O bonds and, most powerfully, C-H bonds – that can take up heat. The layer of greenhouse gases near the Earth's surface traps heat like a blanket.

It was the Swedish scientist Arrhenius who realised in the 1890s that man's activities were slowly – at that time – adding more molecules of, in particular, CO_2 to the atmosphere. Each additional molecule would absorb a bit more heat, thickening the warm blanket around the Earth. Over time, the whole planet would gradually warm. Arrhenius advanced this theory in a book, titled *Worlds in the Making* when it was translated into English in 1908. So public knowledge of the theory behind climate change is at least 100 years old. (Ironically, records show it was at about the same time that the warming trend we are still on began, unbeknownst to Arrhenius or anyone else.)

Fortunately for us, each of the greenhouse gases is at least partially self-regulating:

- Excess water vapour settles as dew or falls as rain, sleet or snow. (But, once warming kicks in, the warmer air can hold more water vapour, increasing warming.)
- CO_2 is slowly removed from the atmosphere through chemical absorption in the oceans, followed by internment in their depths. Adding more CO_2 to the atmosphere speeds this absorption process up somewhat. More CO_2 also encourages plant growth, which absorbs additional CO_2, on land and sea. But the half of emitted CO_2 that isn't so absorbed stays in the air for many decades, even centuries.
- Methane, the remaining major greenhouse gas by quantity and effect, causes warming 20 times more strongly than CO_2. Luckily it breaks down after an average of a dozen years, leaving behind a less absorbent but longer-lived CO_2 molecule to trouble us further.

So Mother Nature can absorb a fair amount of mischief from, say, human activities without disturbance. As long as there were only a few millions or tens of millions of people clearing land (which releases the CO_2 stored in greenery), raising livestock (which generates CO_2 and methane from animals' digestion), burning wood and other fuels (lots of CO_2) and so on, nothing much changed.

But humans started growing in numbers, and became ever more polluting in the process. (The very technology that allowed our numbers to grow increased average emissions per person.) In about 1800, human population reached 1 billion. And a segment of humanity – mostly in the UK, Europe and North America – started to become industrialised: consuming more meat and burning more and more fossil fuels, for heating and cooling, for cooking, transport and manufacturing.

As our numbers grew, the impact of our activities multiplied. Agriculture, seemingly benign and even "natural", is bad for the planet because it requires cleared land and, when practised on an industrial-type scale, artificial fertiliser. An adult needs to eat about three pounds of grain a day to maintain their weight. This requires about half an acre of land to sustain a person.

But eating meat is worse, because the gain for humans is indirect. It takes about seven pounds of grain for a cow to produce a pound of beef. Beef has only about half again the calories of grain. So for each 10% of daily calorie needs a person gets from meat, the acreage needed to support that person almost doubles. People in industrialised countries who get 20% of their calories from red meat are consuming the grain from about three times as much land as a vegetarian.

Not only does this require much more land, water and fertiliser per person. Livestock basically ferment grain in their stomachs before digesting it. This fermentation produces large amounts of methane, mainly emitted by belching and farting, as well as CO_2.

The carbon consumption of transporting feed, the methane emissions from livestock and so on make the disparity even worse. It's been estimated that a pound of beef generates 100 times the greenhouse gas impact of a pound of carrots.

As the number of humans grows, emissions grow too. But emissions grow unequally. Supporting an American emits twenty times as much greenhouse gases as for an Indian; a European, ten times as much; a Chinese, five times. (Some of the Chinese share is to make goods that largely go to the US and Europe, who are thereby offloading their own emissions – plus shipping – to China.)

But wait – where's the proof that all this causes global warming?

We don't have absolute proof, simply because we can't do a controlled experiment. We can't create perfect copies of Earth, then try altering the greenhouse gases differently in each to see what happens.

We do have a solid theory and a good, strong correlation over time between greenhouse gas emissions and temperatures.

There's no competing theory for a cause for climate change. While a contribution from some kind of orbital or geothermal long-term cycle can't be completely ruled out, the scientists who contribute to IPCC reports are now more than 90% certain that global warming and the resulting climate change are caused primarily by human activity.

The actions needed to reduce greenhouse gas emissions are expensive, distracting and painful. Here I differ with what I consider happy talk by many who want to fight climate change, who use words like "easy". Raising the price of home heating during the winter by a "modest" amount – say, 10% – is a big hit for a lot of home budgets. Many people will cut back on heating and a few may freeze to death. Hardly "easy" for them.

These actions, though, are necessary to stop climate change, which will otherwise kill many more. There is more than enough commonsense and scientific evidence as to its reality, and more than enough scientific certainty as to the cause.

Action becomes far more urgent – and inaction far more inexcusable – if climate change has already run away from us, as I assert throughout this book.

Getting people to understand this and act on it, though, will be maddeningly difficult. Not least, ironically, because of the selfsame, wilfully happy – and Nobel Peace Prize-winning – climate change warriors who convinced most of us that climate change is real, as I'll describe in the next chapter.

How emissions hang around

Comparing climate change to other pollution problems is helpful, but not entirely so. Many pollutants leave the environment pretty quickly. Soot, nitrous oxides and most other components of smog, the best-known pollutant, wash out of the air with the next good rain.

Controlling smog in the Western world was made much simpler by this fact. As hard as it was to cut emissions – less polluting fuels had to be used, cars had to be fitted with catalytic converters, factory smokestacks with scrubbers, and on and on – the benefits kicked in quickly. If output of smog's components was cut 5 percent a year, the amount of smog in the air dropped 5 percent a year. (In fact, up to a point smog is invisible, so cuts made visible smog drop even more.) The benefits were obvious and immediate.

Emissions of excess greenhouse gases are much different. There is no immediate visible effect or smell to warn us they're there – to add a sensory element to what our intellects tell us is a bad thing. And they don't wash out of the skies with the next good rain – they stay for decades, even centuries.

The persistence of CO_2, in particular, has huge implications for the environment. It's very hard for people to cut emissions. Yet each bit of CO_2 we add immediately increases the level in the atmosphere; after about half is absorbed by greenery and the seas, much of the remaining half stays for a century or more. It's like an elevator with no "down" button.

As CO_2 and damage caused by climate change increase, the processes that today remove half of CO_2 may also take a turn for the worse. Absorption by plants may soon be overrun by stress on large swathes of the Earth's plant life from excessive heat and drier conditions. Deforestation and forest fires emit CO_2 immediately and take dense greenery out of the picture for many years, in some cases permanently.

The seas may also be moving toward saturation as they absorb ever more CO_2, becoming worryingly more acidic in the process. Warming of the seas also reduces their absorptive capacity, and strong winds resulting from more frequent extreme weather conditions can expose carbon-laden deeper waters to the air, releasing their stored CO_2.

If the free pass that has seen half our CO_2 emissions taken off our hands, so to speak, is ending, atmospheric rises may accelerate faster than even our own rapidly increasing emissions levels would, on their own, merit.

The agony and the enthalpy

One of the hardest aspects of climate change to understand is the gradual impact of warming. Chief among these is the idea that we are already, as of today, committed not only to the warming of 0.8°C that has already occurred, but to an additional warming of perhaps 0.6°C that would occur even if emissions and deforestation stopped tomorrow, and no carbon sinks in nature tipped.

How can this be? I haven't seen a good, succinct explanation, so I'll try to provide one. First is simply the "additional blanket effect" – if you add a blanket to your bed, it takes time for the blanket, and the air trapped beneath it, to get warm. With the CO_2 we've already added to the atmosphere, it takes time for these molecules to trap additional heat energy, and for these energised molecules to have their full effect on the atmosphere, land and sea around them as various air and ocean currents move heat around the planet.

Another reason is the deeper ocean's heat absorbency. We are rightly concerned about, and measure, warming in the zones we live in and interact with directly – the lowest levels of the atmosphere, the uppermost layer of the ocean and the surface of the ground (with its changing snow and ice cover). This is the area in which we measure the 0.8°C of warming so far that we refer to, and the 0.6°C we're still due from the same emissions and deforestation.

But much warming initially affects deeper levels of the ocean, and is only gradually expressed in the narrow slice where we measure warming. For instance, deeper tropical ocean waters absorb a great deal of heat; when they take the brunt of additional warming, it can take years or decades for that heat to affect the measured zone, usually in temperate and polar regions.

This kind of heat is called "latent heat" or enthalpy, a powerful and poorly understood scientific concept. If I find a better explanation of enthalpy and its specific effect on climate change, I'll add it to this book's site, getridofwarming.org; search there for "enthalpy" to find any such addition.

Kyoto vs. climate change

So how is the world responding to climate change?

The most important response so far is the Kyoto Protocol. Inspired hugely by the IPCC process – see next chapter – Kyoto is an international agreement intended to reduce greenhouse gas emissions and thereby limit climate change.

The Kyoto Protocol is based on the Montreal Protocol, the most successful international pollution control effort ever. Generally, pollution control *laws* – at the national, state/provincial and local levels, where people vote or otherwise make themselves heard – are successful, while pollution control *treaties* between nations – which theoretically have the force of law, but are very hard indeed to enforce – are unsuccessful.

In sharp distinction to this general rule, the Montreal Protocol is the tool by which the world reduced, and is now on track to eliminate the danger of CFCs – chlorofluorocarbons, also known by the trade name Freon – used as aerosol propellants and refrigerants.

The 1980s saw an explosion in the use of CFCs. CFC-propelled aerosol hairsprays and reasonably priced automobile air conditioners are just two of the products CFCs made possible.

Unfortunately, CFCs also destroyed stratospheric ozone – the ozone layer between about 20 and 50 miles up that protects us from harsh solar radiation. As CFCs accumulated in the stratosphere, they ate into the ozone, causing the infamous "ozone hole" over the Antarctic as well as increases in skin cancer and crop damage worldwide.

Efforts to reduce CFC usage were battled strenuously by many with an interest in their use, or just a sceptical attitude toward environmental protection, regulation and so on, with evidence of damage argued against every step of the way. (Sound familiar?)

Luckily, the world ignored the sceptics and reached agreement under the Montreal Protocol. Led by rich countries that used the most CFCs – in particular the US, with more than half of usage – CFCs were sharply reduced. President Reagan and the first President Bush, both Republicans, agreed with a majority Democratic Congress to pass, ratify and enforce the protocol.

This effort was a huge success, and just in time. It turned out that damage to the ozone layer was greater than previously believed, and

CFCs more damaging – not least as greenhouse gases – than had been realised. The rich countries reduced usage quickly, and poorer countries more gradually reduced their own use, as intended. Today, CFCs are receding as a threat to the ozone – though, ironically, they're still of moderate importance as a greenhouse gas.

But while the Montreal Protocol is a good model, the Kyoto Protocol has been a miserable failure. It was based on the same principles as Montreal: set an easy initial target; get rich countries to go first; toughen the targets; use the lessons learned so poorer countries can then follow at somewhat reduced cost and uncertainty, buying in technology initially developed by rich countries – which thus make back part of their investments.

For Kyoto, letting the poor countries wait was controversial; though the huge majority of greenhouse gases in the atmosphere were put there by the US and Europe, then-current greenhouse gas emissions were more evenly spread than had been the case with CFCs.

With this in mind, the US – supported by erstwhile ally Australia – refused to go along. In 1999, the US Senate passed a "sense of the Senate" resolution against the Kyoto Protocol by 99-0. Then-Vice President Al Gore signed the treaty, but President George W. Bush formally renounced it shortly after taking office.

And so God, it seemed – or at least a changing climate, which is pretty much indistinguishable in some respects – smote them. The US soon endured Hurricane Katrina, which caused huge damage to New Orleans and to Bush junior's reputation. More intense top-end storms are one of the mainstream – though, again, energetically disputed – predictions for climate change.

Southeastern Australia, where most of its people live, went into a long-term drought that finally helped chase Prime Minister John Howard from office. And the state of Victoria recently suffered its hottest temperatures and deadliest brushfires ever. More frequent and more intense droughts, especially in marginal areas such as southeastern Australia (and the southwestern US, and southern Europe), are another climate change prediction.

Prime Minister Kevin Rudd, Howard's successor, quickly signed the Kyoto Protocol on taking office – as, symbolically, dozens of US states and cities have done – and saw to its ratification, leaving the US isolated.

But why did the US and Australia, alone among major countries, refuse to sign Kyoto when it mattered? The reasons go beyond narrow political considerations and shed light on the future of the fight against climate change.

The US and Australia are two of the world's largest CO_2 emitters. Australia, with a mere 21 million people, emitted 417 million tons of CO_2 in 2006, according to UN figures. That's almost exactly 20 tons of CO_2 per person. The US, with 303 million people – 15 times as many – emitted 5.9 billion tons of CO_2 – nearly 15 times as much, or about 19 and a half tons per person.

Western Europe, every bit as "civilised" as the US or Australia – more so, Western Europeans would sniff – emits about half as much CO_2 per person. What's going on?

This is painful for me, as I'm American and spent several years living in Australia as a child, so I'm talking about my home and one of my favourite countries. But America is home to, and Australia is perhaps the world's most enthusiastic follow-on adopter of, what can be called the American Dream lifestyle, the "supersize me" lifestyle, or any of a number of other names. (The UK, where I live now, is only a bit behind on many of the same tendencies I describe here.)

This lifestyle includes a middle class that mostly lives in big homes placed far apart; extensive car ownership; a big and powerful economy; a big and powerful military; and lots and lots of energy – therefore fossil fuel, in today's world – usage. Economic success allows lifestyle excess, relative to what the planet can support; military power supports access to oil and other resources.

This has all sorts of side effects, some which most people would see as positive, others negative. Despite weak social safety nets, few people starve in either country, and only the poor have to live in homes as small as – or use public transport as much as – the European norm. But the US and Australia have also been successive leaders in obesity and all the health impacts it causes, with far too many Big Macs in one and too many shrimps on the barbie (and tinnies of beer in the reefer) in the other.

Much of daily life in both countries is powered by coal, the most CO_2-intensive and cheapest of the major fossil fuels. Australia gets about 80% of its electricity from coal-fired power plants and is the world's fourth-largest producer; the US gets about 50%, and is the second largest producer of coal in the world.

In fact, the US has been described as a "petro-state" – the world's third largest oil producer; second in coal after China; and third in natural gas. The fossil fuel industry is more influential in the US than in any other Western country – except Australia, where the huge coal industry makes nearly all the running.

Western Europe, on the other hand, has smaller coal reserves, is more densely populated, has smaller homes, has more mass transit, fewer and smaller cars, somewhat less wealth and much less military power. Energy production and use is more highly regulated and much more highly taxed. Air conditioning is used much less. The net effect is an amazing 50% reduction in greenhouse gas emissions per person compared to the US and Australia.

So the fundamental driver of America's and Australia's lack of support for Kyoto was self-interest. Both countries would have to change immensely to quickly move to even European levels of greenhouse gas emissions, let alone the even lower levels needed to actually lessen climate change impacts. People, individually or in groups, try to avoid games they can't win. Bush and Howard, respectively, were just unusually pugnacious advocates of positions that reflected the majority opinion in and core economic interests of their countries – though it was their own decision to make alternately aggressive and dismissive statements on the issue of climate change that will justifiably haunt them when the history of this period is written.

For the US, in particular, the "wedge argument" – a legitimate concern that was used as a tool to divide and conquer the opposition – against Kyoto was the non-participation of developing nations such as China, which were and are growing as contributors of greenhouse gases. Howard was more generally derisive of the whole argument and process, with his dramatic statements taking some heat off Bush.

In the long term, of course, all nations, including developing countries, need to participate in emissions reductions efforts. But, as with the Montreal Protocol, the richer countries – and the biggest emitters per person and biggest contributors to the problem over time – have to "jump first" to make a solution possible. So the China argument, when used as a "stopper" at this particular point in time, was more a red herring than a sensible objection.

Bush and Howard had a personal partnership that allowed them to act as a tag team on the issue. And Bush in particular was willing to attack not just scientific conclusions – hardly a first for a politician – but the scientific process itself, which damages the infrastructure humanity needs to deal with other issues as well as climate change.

President Obama and Prime Minister Rudd, their successors, are more scientifically oriented and more open to the need to combat climate change. This has lessened the pressure on their countries. But neither is proposing anything like what's required to move their outsize emissions close to European norms of about 10 tons of carbon per person per year – let alone the true global average, which is about 4.5 tons. Instead, they each endorse gradual percentage cuts – more gradual than what other developed countries agreed to under Kyoto.

Indeed, the two countries can be seen to have succeeded in an extensive "bait and switch" – first putting forth leaders who attracted worldwide opprobrium on the issue, then gaining favour via new leaders whose friendlier approaches mask positions that are still largely self-serving and that avoid needed fundamental change. As with other countries, the US and Australia can only be expected to make further changes in their position when circumstances change significantly, or when they see big advantages in taking a leadership position, or big problems in not at least following along. Some of those pieces seem to be lining up in both countries, especially in the US.

It's still massively important that Kyoto has failed; this and, I would argue, the problems with IPCC reports described in Chapter 3, have cost the world at least a crucial ten years in responding to climate change.

China, which was a signatory to Kyoto but only required to report emissions, not reduce them, has, through these reports, revealed emissions that were shockingly higher than anyone (perhaps including the Chinese authorities themselves) had believed – and which proceeded to surpass those of the US in 2007.

With China now the leading greenhouse gas emitter in the world, a rich-nations-first approach may no longer be tenable, though expecting developing countries to take on leadership makes no sense either. And countries that have bravely battled to reduce emissions despite the lack of support from others, such as much of Europe and Japan, are now threatening to renege in the face of the credit crunch and global recession.

The low initial targets set under Kyoto, meant to serve as an easy step up on the way to a more serious effort, have been made to seem like hurdles set impossibly high.

Not only has a decade or more of irretrievable time been wasted; having the Kyoto process fail is in many ways worse than never having tried at all. It will now be extremely difficult to convince people worldwide to try the same thing again. The ability to use the Montreal Protocol as a model has been seriously damaged.

It may seem inappropriate for me to bemoan the collapse of Kyoto when it was designed to address a challenge much smaller than what I believe humanity actually faces. But it would have been far easier to scale up a successful Kyoto process – even to a problem many times larger than the previously understood size – than to be facing the huge problems in front of us with no proven model and no, or negative, momentum.

The failure of Kyoto is another way in which the world, and especially younger generations, are, to be blunt, screwed in the climate change crisis. Not only is the problem runaway, with regards to both emissions and the tipping of carbon sinks; the best available mechanism for addressing it has been trashed.

Chapter 3. The 2007 Consensus

> In this chapter
> - Defining the 2007 Consensus
> - Al Gore and *An Inconvenient Truth*
> - The IPCC and its Reports
> - Jim Hansen of NASA
> - The UK's Stern Review
> - James Lovelock and the Gaia hypothesis

"If a million people say a foolish thing, it is still a foolish thing." Anatole France

The argument presented in the previous chapter, that climate change is real and happening now, is largely just a different (and hopefully, to some at least, more convincing) presentation of the case made by others before me. Yet my conclusions about the coming decades, presented later in this book, are much different to mainstream opinion.

If my conclusions are correct – and each day's news further convinces me that they are – how could the current, widely accepted approach be so wrong?

To engineer a paradigm shift – a change from an old way of looking at the world to a new approach – one needs to clearly differentiate between the old and new paradigms.

The new paradigm I'm proposing in this book can be summed up in three words: runaway climate change. To elaborate: what is feared for the future has already happened; nature's carbon sinks have already tipped, and climate change will now change the world even if we halt emissions from human activities and directly human-caused deforestation.

As a corollary, I also propose that greenhouse gas emissions from humans are in the midst of a huge increase, due to roughly treble in this century; that any net decrease in emissions – which would still be adding to the accumulated greenhouse gases in the seas as well as the atmosphere – in the coming decades is very unlikely; and that we're therefore on track to destroy the environment, and quite possibly our civilisation, even if carbon sinks in the natural world weren't gearing up to help pull down the temple around us.

If you look at the evidence presented in this book – even if you just read Al Gore's marvellous *An Inconvenient Truth* with an open mind – you'll see both of these assertions are true.

In fact, I propose to redefine to redefine the term "climate change". We are no longer experiencing *a change in the current climate*; we are experiencing *a change of climates*, a secular ("occurring once in an age") shift from the Holocene climate of the recent past to a much hotter, ice-free climate with extensive desertification which I call Hot Earth. A climate in which, if humanity survives, we will be unaccompanied for millions of years by much in the way of higher plants and animals.

While it still may be possible to stop this shift, it will take a far greater effort than the moderate adjustments proposed to date in response to what I assert is an outmoded definition of climate change.

The new ideas can be likened to a tugboat, and the old ones to a super-tanker. The tugboat can turn the super-tanker around fairly quickly – but only if the super-tanker is pushed in just the right spot, and just hard enough.

The technique will be borrowed from an old boxing maxim: "Move the head, and the body will follow". Let's take a look at the body and head of the old paradigm.

What is the 2007 Consensus?

The 2007 Consensus contains a disturbing set of assertions:
- The Earth is slowly but steadily getting warmer
- The warming is caused by human activity, in particular the emission of greenhouse gases
- There are carbon sinks in nature that, if warmed much further, will "tip" and add to warming, putting further change beyond human influence
- We have just a few decades left in which to reduce emissions and prevent this.

These ideas, and their coalescence into a view of the world, arose gradually, beginning in the late 1970s – when global warming first came into widespread public awareness, leading no less a conservative icon than Margaret Thatcher to call for action on it.

The view that later coalesced as the 2007 Consensus was already outmoded when it first came to light in the 1980s, but the evidence that makes this clear – submarine data on North Pole ice thickness –

was only unearthed and made public by Al Gore, an environmental activist fortuitously holding a top security clearance as Vice President of the United States, in the mid-1990s. The full implications of this data, and the events that have unfolded since, have not been well understood right up to today.

The trouble is, this crucial data was welcomed, if that's the right word, as proof that climate change was occurring (and it is that) – but not examined further for its implications as to the tipping of carbon sinks.

So if the consensus was first mooted in the 1980s, and the crucial evidence supporting most of it was discovered in the 1990s, why call the current mainstream view on climate change the 2007 Consensus?

Because it reached its all-time peak in that year. Gore released his book and movie, both called *An Inconvenient Truth*, in 2006, and in 2007 went on the road promoting them.

The book, movie and presentation brought together ideas Gore had been speaking about, writing about and pushing politically since the 1980s, and his efforts are as good a job of presenting complex scientific ideas to the public as anyone has done, ever.

It was also in 2007 that the Intergovernmental Panel on Climate Change, known far and wide as the IPCC, released its landmark Fourth Report. (Each overall IPCC Report is released in pieces, with the all-important overall Summary coming last.) Because he had been tracking the IPCC for years, Gore's published works included most of what was to appear in the Fourth Report, and he continually updated his presentation as new results came out.

For two decades up to 2007, Dr James Hansen continued to publish crucial work on climate change, testify to Congress and speak out about it. A NASA scientist who has been a leader against climate change since the 1980s, Hansen is famous for making predictions that other scientists might describe as a bit ahead of his data, and for making public policy recommendations that other scientists would consider beyond their remit. (Whereas the IPCC process, described below, is meant to allow science and public policy to interact in a controlled fashion.)

In 2007 the targets some countries agreed to in the Kyoto Protocol began to be implemented. Buoyed by then-expanding economies, European leaders in particular took bold steps toward meeting their treaty commitments. This was despite the then-continuing refusal of the US, by far the largest emitter historically and on a per person basis, to be involved.

And it was in 2007 that then-Sir, now Lord Nicholas Stern's *The Stern Review of the Economics of Climate Change* was published.

Commissioned by then-Chancellor of the Exchequer, now British Prime Minister Gordon Brown, it presented a detailed comparison of the costs of fighting climate change vs. the costs of not fighting it.

Gore's movie won the 2007 Oscar as best documentary, and his book and the DVD of his movie became best-sellers worldwide. In October 2007, Gore and the IPCC shared the Nobel Peace Prize – well-deserved recognition of the many years of hard work that preceded the award.

But it was clear even as all this information came out that the 2007 Consensus was inadequate and overoptimistic. I'll set out the reasons throughout this chapter. And from 2007 to today, new information has emerged that clearly shows the 2007 Consensus to be incomplete and, in important parts of its view of both the recent past and the future, incorrect.

The most dramatic news since the 2007 Consensus coalesced has been the melting of the North Pole's ice cap to record lows – the smallest size ever in 2007, and a size almost as small, with the first recorded opening of the Northern and Northwest Passages, in 2008. Scientists now accept as inevitable that there will be no more permanent ice at the North Pole in one to two decades.

Even more importantly, a recent trend of greenhouse gas emissions being much higher than had previously been the case, and increasing fast, continued. (The increases were predictable to anyone who thought through the implications of economic growth in the developing world, in particular China and India – which several proponents of the 2007 Consensus were extraordinarily well equipped to do.) Gore's book, published in late 2006, showed emissions from China as half those of the US; yet total Chinese emissions passed those of the US in 2007, though still well below European and, doubly so, US levels on a per-person basis.

And in 2008, the global credit crunch hit, marking the beginning of what looks to be a years-long recession. At this writing, governments everywhere – in particular several governments in Europe, so recently leaders in climate change action – are threatening to pull back on their emissions commitments in deference to economic pressures.

The underlying problem in the 2007 consensus is over-optimism; or perhaps insufficient pessimism is a better description. Let's take a look at the three most important publications involved – Gore's book, the IPCC's Fourth Report, and the Stern Review – as well as the work of the important independent voice, Dr James Hansen, and the workings of the Kyoto Protocol – to demonstrate how this makes itself apparent.

Gore's inconvenient truths

Al Gore's *An Inconvenient Truth* – whether delivered as a book, a movie, or a live, in-person presentation – is powerful, impassioned and cogent.

For a mainstream audience that's relatively open-minded – that is to say, most people – it's very convincing, and Gore has indeed convinced many millions of people that climate change is real.

But it doesn't work for everyone. Gore throws together what I've discussed in separate categories above as commonsense evidence, scientific evidence and scientific theory, along with personal anecdotes; those who need a more step-by-step presentation might be left unconvinced. Throwing everything in the pot also has the effect of being absolutist – it's hard to challenge a part of it without seeming to be contradicting the whole.

But the greater problem isn't that his evidence doesn't support his conclusions; it's that his evidence supports stronger conclusions than he draws. Gore's works make a very good case for runaway climate change. There's really nothing to stop all the frightening lines he shows from continuing to move sharply up and to the right.

Gore makes such a good case for runaway climate change that he himself tries to refute it. In a list of misconceptions at the end of his book, his fifth question to himself is, "So does that mean it's too late to do anything?" He answers "No" – then lists things one can do. This is nice, but a bit beside the point. Of course people can do things; the important question is whether it's too late for the actions described to have much effect.

Why is Gore so optimistic? I think that he, along with Dr James Hansen and much of the IPCC, set out in the 1980s to try to save the world from itself. Such an effort takes boundless optimism and great faith. That optimistic attitude and faith have survived an awful lot of ongoing public neglect on the one hand and new, frightening information on the other; perhaps to a fault.

Being a politician first, a populariser second and a gentleman scientist third, Gore is also a bit too willing to tailor his message to suit his audience. He himself tells the story of how he likened mankind's situation to that of a frog in a pan of water.

If a frog is placed in water that's too hot at the beginning, the frog jumps straight out. But if the water is cool when the frog jumps in, then is gradually heated, the frog will be lulled into remaining until he's – rescued. Gore used to say "boiled", but it upset some members of his audience too much.

Al Gore has performed heroic services for humanity. The most stellar single example is when, as Vice President, he saw the above-mentioned US Navy submarine soundings showing the ice thinning at the North Pole. Recognizing it as absolutely crucial evidence of global warming, he had it cleaned up to remove "sources and methods" (key intelligence information, like just where America's nuclear submarines were on specific dates) and released into the public domain. This information continues to be crucial in our understanding of the roots of climate change.

But the world might be a different place today if Gore had then asked his scientist friends two questions. First – I can hear Gore's drawl in my head – "How long before *all* of the ice melts?" Then, "Having lost all that reflective ice, won't the Earth be forever warmer – and in runaway climate change – once that happens?"

But Gore, for all he's done for humanity, was not going to be the one to boil that frog.

The broken IPCC process

The Intergovernmental Panel on Climate Change is a marvel. Gathering together the work of more than 3000 researchers, the IPCC is the largest working group of scientists ever assembled.

But the IPCC is not only a scientific exercise. It's an inter*governmental* body, after all. It includes numerous policymakers.

These policymakers play a troubling role. They have input into every line of the reports the IPCC puts out, which too often means that the archconservatives among them – at different times from the US, Chinese, Russian and Saudi contingents (all fossil fuel giants) and others – challenge any strong statements.

Worse, the policymakers write the overall Summary for the Reports the IPCC issues every six or so years. In earlier Reports, this meant the Summary was a hotly contested document. For 2007, the arguing was reported to have been reduced by the policymakers ignoring the scientists to an unprecedented degree, issuing a Summary that made up for in incomprehensibility anything it lacked in optimism.

(Go online and see for yourself, using the references in Appendix C. In contrast, Gore's work, and even Hansen's scientific papers, are far more understandable.)

One unfortunate example of both optimism and obfuscation is echoed in many recent IPCC documents, other scientific reports, government analyses and the popular press. Prominent in the 2007 Report are a set of scenarios for climate change worked out in cooperation between scientists and policymakers on the IPCC back in 2000. The Scenarios attempt to capture different possibilities for how global society could evolve to reduce its emissions, but they appear fanciful – and distracting – compared to the harsh realities of emissions, which have recently tended to meet or even exceed a simple continuation of current rates of growth.

Figure 3-1 shows a representative set of IPCC emissions scenarios, in a version of a figure that has received repeated global publicity and use. The Scenarios were specifically developed to *exclude* any focussed efforts to reduce emissions; that is, they showed what might happen if no efforts are made expressly to reduce emissions. Yet most of them include considerable reforestation later in the century and, in almost all scenarios, big and unjustified drops in emissions below current trends – some in the near term, some a few decades from now, or both.

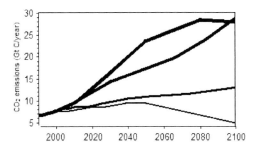

Figure 3-1. Four IPCC emissions scenarios

Why is this so wrong? Because, since the world's economies are heavily dependent on fossil fuels, economic growth causes increases in emissions. The world's economy is projected to grow about 3% a year throughout this century – doubling in size four times. Absent dramatic (and targeted) action, emissions would tend to do the same. So there's no way for emissions to actually drop short of targeted action or economic disaster. Yet the Scenarios exclude the first *a priori*, and aren't based on any prediction of the second – yet they somehow include sharp drops in emissions.

Not only are the Scenarios – there are over 40 in total – fundamentally wrong, they could hardly be more confusing if they were designed to be. Here is a description of just one "scenario family" from the Special Report: Emission Scenarios, published in 2000, which first set them out:

> "The B1 storyline and scenario family describes a convergent world with the same global population that peaks in mid-century and declines thereafter, as in the A1 storyline, but with rapid changes in

economic structures toward a service and information economy, with reductions in material intensity, and the introduction of clean and resource-efficient technologies. The emphasis is on global solutions to economic, social, and environmental sustainability, including improved equity, but without additional climate initiatives."

What does this even mean? What is a "convergent world"? But the money words are at the end: "without additional climate initiatives." In the real world, without additional climate initiatives (or disaster), emissions will rise and rise, at least until everyone is well off. Yet the B1 family, like other IPCC Scenarios, shows them dropping.

The effect of the Scenarios has been awful. People, not unreasonably, look at charts like Figure 3-1, figure that the high-end predictions are a worst case scenario, a reasonable projection is somewhere in the middle – and that, with effort, we can achieve one of the lower-end projections. The truth is that a "reasonable projection", using information known at the time of the 2007 IPCC reports, was for steady growth above even the highest of the Scenarios.

Here, as just one of many available examples, is a quote from Andrew Sullivan, a well-known writer for *The Atlantic Monthly* and many other publications, from the London *Sunday Times*, 5th April, 2009: "...accepting the middle range of options that the IPCC has set out, doing nothing at this point may be the least worst option." Though very intelligent, Sullivan has read the Scenarios as options – not projections – and thinks the middle range is presented as the most likely future possibility.

Yet emissions since 2000 have, entirely predictably, been higher than even the highest-end of the Scenarios, and increasing steadily – not tailing off dramatically at some points, as all the Scenarios do, which would be a first in recent human history. And further, efforts to reduce emissions – which are specifically excluded from the Scenarios – have so far shown no effect at all.

The "Sullivan effect" occurs over and over. The IPCC Scenarios are used again and again, all over the world, as a basis to project the impact of climate change on various important issues, from food production to desertification. Yet they're rubbish. A great deal of humanity's response to climate change, in policy and in popular opinion, is based on this ridiculously rosy set of scenarios.

What should the IPCC have used? Routinely, in science, business or any other rationally based endeavour, the initial estimate of the future is based on a projection from the past. Then you adjust it for factors that may change the trend, carefully spelling them out. And you then acknowledge that wildcards – "black swans", as they've come to be known – could upend your projections, and any plans based on them.

And, when you make a projection, and then things change, you adjust your projection. In the case of greenhouse gas emissions, increases have soared – from just over 1% a year in the 1990s (a doubling of total emissions every 65 years) to increases of just over 3% a year from 2000 on (a doubling every 23 years). With this kind of shift, you would study the situation to understand the causes of the change. (In the case of emissions, rapid economic growth in developing countries, particularly China and India.) And, if you are really wrong – as the IPCC was in its 2001 Scenarios – you go back and study your own process to understand how you missed the mark so badly. The IPCC is doing none of this. One can only hope it will.

Figure 3-2 shows the selected IPCC projections above, plus a projection of emissions growth using the experience of the years from 2000 through 2007. (This projection does not reflect the slowdown of 2008 and the current global recession, which will probably slow the rise only slightly – as there is still some economic growth in the world even at this writing, particularly in emissions-intensive China.)

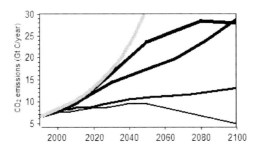

Figure 3-2. Four IPCC emissions scenarios plus curve with continued 3% growth

The new curve shown in Figure 3-2 far exceeds anything the IPCC has projected. It implies much greater temperature increases, especially in the second half of the century – and the rapid tipping of carbon sinks in nature, as described in Chapter 7, even if they had no effect on tipping one another.

There are reasons to believe that emissions will drop, even in the absence of targeted efforts, *near the end of this century*, when the trends driving them upward – industrialisation spreading throughout global populations and population increases – reach saturation point and begin a gradual decline, respectively. (And not unrelatedly.) But such declines will be too late to prevent both huge direct damage from emissions and runaway climate change.

The 2000 IPCC Scenarios never had a firm foundation. Their promulgation in the first place – and their continued use after events had demonstrated how far off they were – is the second great mistake in modern climate science. (The first being the satellite data errors described in Chapter 2.) The credulity of the press and policymakers in reproducing and reusing them without question is astounding; their continued use today, without modification or even caution from the IPCC, only continues to compound the original errors.

Along with the dependence on baseless Scenarios, the editorial process used to create IPCC Reports, and in particular the crucial Summary for Policymakers, was also subject to problems and even manipulation. The IPCC process had worked, after a fashion, for the previous Reports. The very contortions the participants went through to get agreement made their conclusions, though quite conservative, usefully difficult to dispute.

But many, participants and non-participants, do dispute the 2007 report. Some outsiders asserted that the IPCC's conclusions about sea level rises, for instance, projecting only a half a metre rise by the turn of the century – *less* than the prediction in the previous, 2001 Report – ignored definitive recent work projecting larger rises. Projections of potential rises due to ice melting were simply excluded – because, ironically, they were changing (getting larger) too fast. The IPCC acknowledged it was excluding the most important data – then gave projections specific down to the millimetre anyway.

Now there are credible projections of up to a 1m rise – enough to displace tens of millions of people, render great swathes of the world's prime real estate valueless, and put much of London, Washington, DC and southern Florida, among other vital locations, underwater – *by 2050.*

In addition to the continued use of the baseless 2000 Scenarios, and the sea level fiasco, many other changes were made that softened the impact of the Report and implied far less urgency than is in fact required. Individual government representatives filibustered, arguing for hours over a single sentence – and generally getting their way. A few brave scientists involved in the IPCC process even leaked secret minutes to expose how baldly they were overruled. And events soon overtook the panel's all too tentative conclusions, though they continue to be quoted and used.

The next IPCC Report has been put off a year to 2014. It promises to become a major battleground between different factions among both scientists and governments – a battle which, clearly, needs to happen. The world is very much dependent on the IPCC process to know what to think about climate change, but the process in its current form may have outlived its usefulness.

Hansen's Alternative Scenario

Jim Hansen is perhaps the first, and one of the true heroes of the movement to get climate change recognised as a global problem and acted on. Despite the risk to his position at NASA from bosses and top politicians opposed to recognition of climate change, Hansen has been speaking out, in scientific papers, in US Congressional hearings and in public, since the 1980s.

Even Dr Hansen's work is not immune from political interference. At least once, the overview portion of his Congressional testimony was changed by the (George W.) Bush White House. Fortunately, the change conflicted with his main narrative so egregiously that the difference was noticed; Hansen admitted the change during his testimony. Thus, the change caused more embarrassment than benefit for his bosses.

Dr Hansen is unusual in that he not only "does" science, holding down a top position and writing well-regarded and much-cited scientific papers. He also speaks out and writes on solutions to the problems he researches, including policy recommendations within his actual scientific papers. Most scientists refuse to do this, believing that it conflicts with a more narrow responsibility to get at underlying scientific truths.

Hansen is the most radical of the mainstream voices, claiming in 2006 that humanity only had ten years from that date to identify and start implementing low-carbon infrastructure, or the battle against global warming would be lost – a widely-quoted claim.

So why, given his courage and honesty, is Dr Hansen not recognising, researching and speaking out about climate change having already gone runaway?

The answer, I believe, is similar to the problem afflicting the work of Gore and the IPCC. Hansen has sketched out and advocates the Alternative Scenario, a programme for reducing emissions and the resulting climate change.

He suggests that climate change is as much the fault of lesser greenhouse gases and soot as of CO_2, and that cutting them is a cheap win that can buy time while other, more extensive, less-polluting infrastructure is installed.

In pushing for this policy approach, though, Hansen seems to have held back on asking penetrating questions about the underlying situation. It seems as though he needs the window to be open so he can squeeze his solution through it.

Hansen is also moving the goalposts that he asks people and policymakers to aim for. After signing on to other, higher targets, Hansen has recently suggested that a target of as low as 350ppm for greenhouse gas levels – below today's 380ppm – may be needed to prevent disastrous consequences. Which implies that climate change is already in a runaway state, but without actually saying so.

Starting from first principles, as I'll describe in Chapter 9, could have told him the right answer on emissions 20 years ago – a zero-based approach targeting 280ppm – but even a very strong and conscientious advocate like Hansen continues to want to tinker with humanity's future.

It may be – as a scientific purist might suggest – that Hansen's public policy advocacy is obscuring his otherwise clear and penetrating vision on the underlying scientific problem. Those of us concerned about climate change can only hope that he renews his focus on diagnosis ahead of trying to identify and promulgate any specific, and possibly insufficient, cure.

A none too Stern review

Along with Gore's optimism, the hijacking of the IPCC process by policymakers, and Hansen's Alternative Scenario, the *Stern Review of the Economics of Climate Change* is another missed opportunity. The Stern Review surveyed the state of climate change – and the economics of responding to it vs. trying to ignore it.

It should have been a huge accomplishment – and in some ways it is. The *Stern Review* brings together a vast array of information, including specific projections for many countries and a thorough breakdown of the carbon footprint of more than 100 economic sectors in the UK. And its economic analysis is admirably detailed.

There are problems, though. The first is that then-Sir Nicholas, despite having been chief economist for the World Bank, somehow managed to ignore the growth of China and the developing world in his analysis. As we'll see later in this book, China is already the leading emitter, and increasing emissions frighteningly fast. Yet Stern writes as if controlling emissions is almost entirely a Western problem, and will remain so for decades to come. He discusses tweaks that might be appropriate for the West without addressing the huge, almost immediate step changes that would be needed to support emissions-reduced development by the East.

The second problem is that Stern does an admirable job of calculating the plusses and minuses of various climate change scenarios – but he ignores any value for human life in his calculations. In Stern's figures, if a billion people die but the world gets 15 per cent richer while ignoring their fate, the world comes out ahead. Given that world population is expected to grow to more than 9 billion people by 2050, with most of the growth projected in exactly the areas that will be hardest hit by climate change, this is an untenable, as well as amoral, exclusion.

The third issue is that Stern almost ignores the possibility of runaway climate change. He clearly states that a temperature rise of more than 2°C poses unacceptable risks – but then lets his economic formulas run over scenarios with temperature increases of 5°C and 6°C, in which they would clearly no longer apply, as the Earth suffered from major disasters and careened toward worse ones.

The *Review* puts economics in the driver's seat, ignoring that it's the study of a human activity, not a fundamental science like math or physics. If the people who engage in economic activity are suffering too much, too frightened or starved to function, in open revolt, or prematurely dead, standard economic models no longer apply.

What was needed, which Stern did not provide, was an economics of limits – a model for using economics to help sharply limit temperature increases and minimise loss of life while halting emissions and climate change in their tracks. Stern, being "only" an economist, would still have had to operate within the flawed scientific framework of the optimistic 2007 Consensus – but a model based on limits could have been adapted to the harsher realities of any newer paradigm.

As it stands, if a Nobel Prize for Economics is ever given for work related to climate change, it may well go to the economist who comes up with such a model, rather than to Lord Stern.

More false hope

There are many other factors giving us hope as to whether climate change is at least on the way to being dealt with effectively. What I'm saying here is that, at this point, it's false hope.

As mentioned above, the implementation of the Kyoto Protocol has developed in perhaps the worst possible way, as an effort that was initially strong and credible, based on a proven model, but that has failed.

A lot of smaller steps have similarly encouraged an optimistic feeling. Companies have "gone green" – the best example being the department store Marks & Spencer in the UK, with its all-encompassing Plan A – "because there is no Plan B", as the company puts it. American states and cities have symbolically signed onto the Kyoto Protocol in the absence of leadership from Washington, and some – notably the state of California – have begun passing and even implementing tough regulations and strong incentives toward reducing climate change. At this writing, even the US President, his administration and Congress are moving to slow the rate at which America pours greenhouse gas emissions into the environment.

These are marvellous and admirable steps. The problem is that they represent only first steps on a long, long journey, yet one that must somehow be completed very quickly. The 2007 Consensus promotes over-optimism about how much progress these steps represent.

Even the long-term news cycle encourages optimism. There was a huge press focus on climate change from the release of Gore's *An Inconvenient Truth* in 2006, peaking in 2007, through to the point where Gore and the IPCC won the Nobel Peace Prize late in that same year. But then, with little new "news", the whole story seemed to run out of steam.

Moving into 2008, the US Presidential election, the worldwide credit crunch and the first 100 days of President Obama's administration grabbed the headlines – accompanied by ever-increasing penetration of celebrity "news" into mainstream media and layoffs of thousands of experienced reporters worldwide. In a poll in early 2009, the percentage of people who perceived that climate change was being covered as a "major" story dropped from more than half to only a quarter.

President Obama, by the way, "gets it" – at least within the limits of the 2007 Consensus' view of the situation. He has made battling climate change one of his top three priorities, along with the economy and health care. This is at a time when it doesn't even make most Americans' top five. President Obama, his administration, and the Democratic leadership in Congress are all about as far out in front of their constituents as elected officials and their appointees can afford to get.

So this quieter period is a good point at which to assess the results of the 2007 Consensus and of its proponents' efforts. People around the world have been made far more aware of climate change and its causes. It's rocketed up the charts, so to speak, as a concern among people around the world, even if the level of concern has since settled a bit.

But people have been given the belief that tackling climate change is do-able, then seen people in power fail to meet (or, in the case of

Australia and the US until very recently, to even seriously consider) the targets set out as the minimum for effective action.

All along, runaway climate change has been held out as a bogeyman – something very bad, which it is, that will come get you, but only if you don't do the right thing. Which is simply wrong because, as I'll show in the next several chapters, runaway climate change is already here – doubly so, as there are two runaway processes: both human emissions and the tipping of carbon sinks are beyond any short-term efforts at taming them.

The Nobel committee was right to award its coveted Peace Prize to Gore and, for its efforts over more than 20 years, the IPCC. But all those involved need to see that they have not found "the answer". Instead, their efforts have flaws, and have only brought the world one crucial, but small step toward what will need to be a giant leap if we are to survive the disaster we have unintentionally created.

A Gaia future?

The 2007 Consensus did have one strong dissenting voice: James Lovelock, creator of the famous Gaia hypothesis.

The Gaia hypothesis has a weak and a strong form. The weak form simply points out, quite reasonably, that all creatures come into life in an environment that has evolved over time. Each species and many elements of the natural environment affect and are affected by each other. This is a simple principle, but mind-bendingly complex to apply, and Lovelock deserves credit for bringing it front and centre.

The strong form of the Gaia hypothesis, which Lovelock sometimes claims to be a caricature of his views, holds that there is something like a planetary consciousness – and, in some readings, that it doesn't like people very much.

Lovelock will always have detractors among those who haven't forgotten that he was a conscientious objector in World War II Britain, converting to become a Quaker in time to avoid service. He says he then changed his mind about serving when he learned more about Hitler – not a common reversal among true Quakers – but was then not allowed to leave his research position to fight.

Yet the Gaia hypothesis – at least its more tentative form – is a real contribution. Unfortunately, Lovelock turned to hyperbole to promote his 2006 book, *The Revenge of Gaia* – a title that seems to attribute both emotion and intelligence to the planet. The subtitle, which begins *Why the earth is fighting back...*, further advances the theme of Earth as spurned goddess.

In this book, Lovelock presciently observed that climate change was more likely to go out of control than not, but made dramatic claims that only a billion humans will remain by the end of the century. While widespread death is not out of the question, it's hard to defensibly vouch for 6 billion deaths over 5 billion, 3 billion, "everybody" or some other number. The spuriously precise number made for eye-catching headlines, though.

Lovelock has described both World War II and possible upcoming climate disasters as "exciting", which ignores the vast amounts of suffering involved for those on the front lines, which Lovelock – in the first instance by design, in the other by the luck of the draw – won't experience. And he described humans as an "irritating species" which climate change is intended, by Gaia, to rid the Earth of, again trying to have both the strong and the weak versions of the Gaia hypothesis at once.

The Revenge of Gaia was insightful, but delivered in a way that made all but Lovelock's fans more angry than concerned. By presenting valid analysis in such a self-servingly dramatic fashion, he made it easy for those with either valid doubts, or a self-serving interest in preventing action, to dismiss him.

Lovelock's new book, *The Vanishing Face of Gaia* (2009), is more moderate in tone, better suited to a truly senior scientist – Lovelock was 89 at the time of publication – delivering a needed message late in life. But the book is steeped in the Gaia hypothesis, with significant points on climate change, which could stand very well on their own, positioned as supporting or extending it instead.

With regard to understanding climate change, the Gaia hypothesis is unnecessary, a violation of Occam's Razor – the useful observation that simpler explanations are usually the best. One can easily reach similar conclusions to Lovelock's without invoking Gaia – and can be led down all sorts of distracting side paths in doing so.

The Gaia hypothesis, especially when used to portray the planet as a vengeful entity attacking humans with the climatological tools at her disposal, simply makes it easy to dismiss valid concerns about the development of climate change and non-2007 Consensus conclusions about how it's likely to play out.

Lovelock had something to sell – his Gaia hypothesis – and he used what was otherwise new and insightful thinking about climate change to sell it, and to reinforce his position as a sort of elder *enfant terrible* of British science. In doing so he made life harder for those who reach somewhat similar conclusions but are not rapt devotees of his overarching theory.

Chapter 4. Why Emissions are Bound to Rise

In this chapter
- The effects of emissions
- Emissions in the 2007 Consensus
- How economic growth produces emissions growth
- Slowing the battle to cut emissions
- The Institution of Mechanical Engineers
- Emissions trends in coming decades

In 1991, in her pomp as British Prime Minister, Margaret Thatcher promised, "I will go on and on". She didn't, of course; under immense public pressure "the vegetables", as the ITV show Spitting Image once called her Cabinet, evolved a backbone and dumped her. The consequences are still with us today.

While humanity has been cutting down and burning trees for millennia, it took the deforesting efforts of a population that had passed one billion people, and the Industrial Revolution with its use of coal and oil, to really send emissions soaring. Will they go on and on?

One of my purposes in this book is to show the deep behavioural and technological roots of deforestation and greenhouse gas emissions so that we can work to develop effective solutions – and not hack one another to bits over ineffective, even destructive ones. (First generation biofuels, anyone?)

Today, deforestation and greenhouse gas emissions have greater momentum than even "Maggie" in her prime. They will continue to rise for many decades to come unless something at least as shocking as vegetables developing backbones occurs.

In this chapter, I describe the incredible momentum emissions have – and show how their rise will naturally slow late this century, but only after achieving an intolerably, even incredibly, high level.

Do remember, as we go through this, that I haven't even gotten to the core of my argument yet – that carbon sinks in nature are tipping and will cause a frightening step change in our climate. In first presenting a realistic view of projected emissions, I'm just cleaning up the mess – I won't say "draining the swamp", we shouldn't be doing that any more – left by previous efforts, in particular the IPCC's, to project future emissions.

Whence emissions

There are many ways to "slice and dice" emissions, but here's a simple breakdown from 2006 for the US, as also shown in Figure 4-1:
- Electricity generation 33%; mostly fossil fuels
- Transport 28% (cars 51%, trucks 35%, ships 5%, planes 4%, rail 3%, bus 2%); nearly all fossil fuels
- Industry 20%; nearly all fossil fuels
- Agriculture 8%; fossil fuel use, plus methane and CO_2 from livestock
- Commercial buildings 6%; mostly heating, powered by natural gas, plus increasing air conditioning powered by electricity
- Residential 5%; same as for commercial buildings.

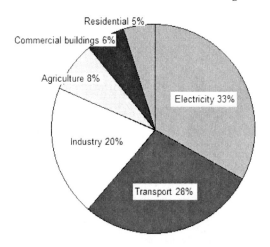

Figure 4-1. US emissions by source

I'm using US numbers because it's the biggest and best-studied economy in the world, at the top in total historical emissions and near the top in current greenhouse gas emissions per person. It also represents the economic, lifestyle and emissions model that much of the developing world seems headed toward. We can't solve climate change without solving emissions in the US. Other sets of figures, such as worldwide totals, mix developed-world apples with developing-world oranges and therefore confuse as much as they illuminate.

In the US, as in most other countries, most electricity is generated by burning fossil fuels – coal, oil, and natural gas. The "dirtiest", coal, is about twice as emitting, per energy generated, as natural gas. Oil is in the middle. A small share of US electricity is generated from renewable sources – nuclear power, hydro-electric power from dams, wind and solar power and biofuels.

Transportation is powered by oil – gasoline and diesel. Industry runs from a mix of fuels, mostly fossil fuels. Agriculture generates most of its greenhouse gases directly from the belching and "tailpipe emissions" of barnyard animals. Commercial and residential buildings' non-electrical emissions are largely from the use of natural gas for heat, with increasing electricity use for air conditioning.

Electricity use is nearly equally shared among industry, commercial buildings and residences. Dividing the electricity share (33%) among its users changes the picture a bit, as also shown in Figure 4-2:
- Industry 29%
- Transport 28%
- Commercial buildings 17%
- Residential 17%
- Agriculture 9%

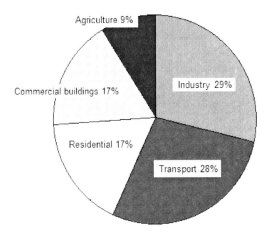

Figure 4-2. US emissions by source, with electricity divided among users

All this leaves out deforestation – emissions caused in the clearing of land – which is roughly 20% of total worldwide emissions, but varies dramatically by country. Some countries have little; in other cases, such as the rapid deforestation, including deliberate fires and associated accidental fires, in Indonesia in 1997-98, a single country's deforestation emissions can constitute a significant proportion of worldwide greenhouse gas emissions in a given year.

Deforestation also reduces current and future capacity for absorbing greenhouse gases. Stopping deforestation is vital – but with more people needing more cropland for food, the onset of biofuels, the developing world paving over more land for cities and cars, plus warming-related increases in insect damage to trees and forest fires, it's hard to see a way to do it.

We'll delve into the details later, but one thing these breakdowns say is that there's no single, easy answer to reducing greenhouse gas emissions. A lot of investment and a lot of change right across society will be needed – if the task is seriously taken on at all.

Why emissions matter

Greenhouse gas emissions matter for two reasons: their direct effect on warming and their ability to trigger the collapse of carbon sinks, which is an indirect, but potentially decisive, contributor to warming.

Emissions and the warming they cause trigger the collapse of carbon sinks in two ways. One is by literally saturating the carbon sinks so they can't take on additional CO_2. (Methane breaks down anyway; less important gases vary.)

There are two carbon sinks that can be saturated: oceans and greenery. There is a point of saturation where the chemical process by which the oceans absorb CO_2 will begin to slow; there is evidence we've reached this point already.

Presented with more CO_2, greenery tends to compensate by growing more extensive and denser, but humans are not allowing this point to be reached – we're destroying greenery, especially the densest (rain forests), quite steadily. And the remaining greenery can be saturated as well; even presented with extra CO_2, limitations on available space, nutrients or water slow or prevent new or additional growth.

So we're reducing the greenery carbon sink, and the remaining extent of it is becoming saturated. Left alone and given many centuries and even millennia, the greenery mix would change and recover, but this process is too slow to help us avoid potentially immense problems in the coming decades.

Beyond direct saturation, emissions cause warming that damages both of the major active carbon sinks. Warmer waters absorb less CO_2; warmer plants suffer stress and stop growing, while becoming more vulnerable to insect damage and fire.

Finally, warming caused by emissions causes melting of the cryosphere, reducing the average amount of snow and ice cover on land and sea, so less sunlight is reflected; more is absorbed to heat the planet. This is a long-term change, already occurring very quickly at the North Pole, with, I believe, decisive consequences.

So emissions matter in several ways. They cause the warming we experience today and, by damaging carbon sinks, lay the groundwork for far more rapid warming tomorrow.

Emissions and the 2007 Consensus

The problem of emissions can be summed up rather simply. While deforestation is a worldwide issue, emissions come almost entirely from industrialised countries – the developed world, or the First World as it was once called, with about 1 billion people. A typical developed country has emissions ranging from about 10 tons of CO_2 per person per year (most of Europe, Japan) to about 20 tons of CO_2 per person per year (the US, Australia, Russia).

The natural course of events for emissions in developed countries is to rise at about half the rate of economic growth (largely due to the fact that half of economic growth relates to lower-carbon-emissions services rather than higher-carbon-emissions "stuff"). Slowing the rate of growth of emissions without slowing economic growth is very, very hard. To actually and truly cut emissions while continuing to support conventional economic growth is nearly impossible to do except as part of a decades-long and global focus, and it would require a wrenching effort even then.

But gradual percentage cuts by the developed world hardly matter while countries in the developing world are multiplying their emissions by two, four or ten times over the coming decades, catching up with the Joneses. The average emissions in the developing world range from about 1 tonne of CO_2 per person per year in the least developed countries up to the 5 tons of CO_2 per person per year in fast-developing China. You can easily see the problem (though the IPCC and other proponents of the 2007 Consensus couldn't); if the 5.5 billion people in the developing world achieve similar lifestyles, and similar emissions, to the 1 billion people in the developed world, total global emissions will treble or quadruple. If the 2.5 billion additional people expected to join humanity by 2050 also join in, emissions could reach five to six times today's. And all of this economic and population growth – and the concomitant emissions growth – is on track to occur in this century, as long as the current economic downturn is overcome in the next few years.

Of course, reductions can be made. But when the underlying trend is for many times today's emissions, any but the most strenuous reduction efforts will only reduce the growth to a slightly smaller multiple – not a cut at all, just a slowing of a very high rate of increase.

And to really stack the deck, emissions in China are quite high compared to their level of economic development or economic growth – partly because they necessarily use older technology, partly because they're taking in the developed world's washing, in the form of doing its manufacturing, and partly because of a tremendous dependence on coal – and at the current rate China will do well to ultimately limit itself to American, never mind European, emissions levels per person.

The 2007 Consensus directly addressed the topic of emissions. The best reflection of the Consensus showed up in the European Union's plans for what's called "20/20/20": a 20% cut in emissions from 1990 levels, and 20% of all energy from renewable sources, by 2020. Following this, politicians have competed to promise either a 50% or an 80% total cut from 1990 levels by 2050.

The US has not come close to matching even the middle-of-the-road European commitment. The incoming Obama administration has recently mooted the idea of cutting emissions to slightly below 1990 levels by 2020 – a considerable cut from the increases since, but well behind the Europeans. And, given the huge volume of goods imported, making sure any American reductions are "real", and not just shifting emissions to suppliers in other countries, will take some detective work.

There isn't yet any discussion of cuts by China, India and other emerging nations, nor a return to 1990 levels – there's been far too much economic growth, more than a doubling in total, with equivalent emissions growth, for that. The discussion for the emerging nations is likely to focus on, first, constraining the growth of emissions, then at some point holding emissions level while continuing to grow economically. (Though this is nearly impossible with manufacturing-led growth rather than services-led growth.) But this discussion has not even started yet, and the developed nations can at best expect the developing ones to eventually converge with their own, currently very high, levels.

Two related questions are crucial to understanding the plans that do exist, such as "20/20/20" and 80% cuts by 2050 in the developed world: are they sufficient? And are they likely to be achieved?

Unfortunately, the answers to these questions are "no" and "no".

The final part of the problem is attitudinal. The 2007 Consensus rightly focussed first on the US and Europe. But then the thinking went all fuzzy.

In Gore's book, the Stern Review and James Hansen's scientific papers, there's a steadfast refusal to take a hard look at the developing world's actual emissions, their likely rate of growth as the underlying economies grow, and what it would take to control them.

There seems to be a tacit belief that there's plenty of time to address the developing world's emissions, and a tacit willingness to let the developing world largely catch up with the West in both economic development and emissions, then cut from there. This is very thoughtful and sensitive, but it doesn't reflect what's needed to avoid disaster.

I assert that runaway climate change has already begun, as I'll explain in Chapter 7. So the chances of avoiding it by the cuts planned, or any other cuts, absent cleaning up excess CO_2 directly, are nil.

Yet the stated goals, while insufficient in depth and coverage, and too late to achieve their goals of containing warming and preventing runaway change, are hugely important.

The world needs to get used to acting in concert and to stringing together successes in the battle against climate change. Emissions and deforestation, being fairly well understood and theoretically, at least, within humanity's control, are the right place to start. Removal of excess CO_2 from the environment then has a far better chance of success.

For even the most willing, and the most able – such as the European countries and, with Obama in office, the Americans – to fail to meet admittedly inadequate goals would be a huge blow to future efforts to combat climate change.

In the discussion below, I leave out cuts in emissions planned by the US, Europe and Japan. This is because only small cuts to 2020 by Europe and Japan are actually committed to.

If further cuts by these countries are agreed beyond 2020, it will make a difference, but only a small one – the real running will be made, for better or worse, by the developing world.

So, in line with other recent analyses (see below), I leave out currently discussed cuts from my projections here. When the hoped-for cuts become more solid, they can be factored in.

How growth is tied to emissions

One of the crucial things that's not well understood about emissions – in particular CO_2 emissions – is just how hard it is to cut them.

There are two links here. Firstly, energy usage tends to grow with economic growth. The economy is a bit like a shark – it's either moving forward (growing) or it's dying.

This is especially true in countries growing in population, which need economic growth higher than population growth to avoid slipping backwards. Most of the world's population growth to 2050, projected at 2.6 billion people, will occur in just three countries: India, China and the US.

Some environmentalists – who, for people who love nature, can be quick to grab the wrong end of a stick – draw a simple conclusion from this: population growth is bad. Not a challenge, which it is, nor even inconvenient, which any parent will agree with them on, but bad: immoral and wrong (by exactly what moral standard is unclear, since most environmentalists are not explicitly religious).

This puts some environmentalists in the uncomfortable position of being dead set against the most deeply felt needs of people in general and the poor, who have the most children, as well as the religious, who tend to value life most highly, in particular. Pejorative terms like "eco-Nazis" resonate all too strongly in the face of positions like this.

With or without population growth, the world is committed to economic growth, not least to help lift up the poor. The link between economic growth and energy use is basically unity in developing countries – one percentage point of economic growth means one percentage point of growth in energy usage and therefore, with today's mix of generating capabilities, in emissions.

The two links are often stated as one: increases in economic growth (which is good) cause increases in greenhouse gas emissions (which are bad). This is true today, but what needs fixing is the gearing in between: energy generation needs to grow to support economic growth; greenhouse gases from energy generation need to drop quickly toward zero.

We already have several nearly zero-emission energy sources: hydro-electric power, nuclear power, solar power, wind power and wave power. Some of these are in quite widespread use in various countries. They all have problems, but those problems – and our energy problems as a whole – are solvable. But we need the will to make the huge investments to tackle them head-on.

And until we do, increases in inputs – more people and economic growth – will continue to cause increases in greenhouse gas emissions. Weakening, then breaking these links will take decades.

It costs a lot of money to generate greenhouse gases – that is, to build power plants, to build car-making factories and so on. Emissions represent years of planning and billions of dollars in investment. So if emissions are rocketing up, there must be powerful forces behind the rise.

With this in mind, Figure 4-3 below may be the scariest graphic you'll see in your life – especially the Total line. It's the curve of greenhouse gas emissions over the last few centuries, with emissions rising from the beginning of the Industrial Revolution in about 1760. What's scariest is the upward curve at the end, which doesn't include the last few years of even sharper rises.

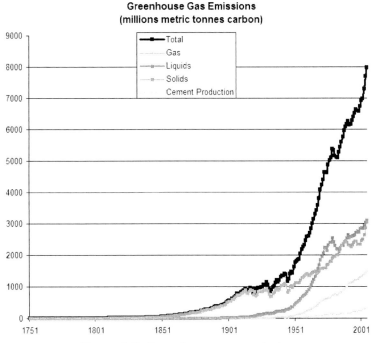

Figure 4-3. Greenhouse gas emissions since industrialisation

Our odds of changing the shape of this curve any time soon are obvious from the curve itself. From the curve, it looks like humanity as a whole, each nation, and every individual human being, have committed to emitting as large a quantity of greenhouse gases as possible.

The explanation for this curve, frightening as it is, is simple. The developed world's emissions are continuing to increase, and increasing parts of the developing world's emissions are rising rapidly as it, well, develops.

There is no inherent reason for this to stop – in fact, the world's economic future is dependent on its continuing, even increasing. China, in particular, is keeping the global economy afloat in the current crunch. The developing world's progress alone means that something like 3% a year growth in emissions is likely to continue for decades.

What does this mean for greenhouse gas levels? You may know the "rule of 72". If you divide 72 by an interest rate, the result is the number of years it will take your investment to double.

If you divide 72 by 3 percent per year, you get 24 years. That means emissions are on track to double every 24 years. So in this century, they're on track to double from today's levels 3 times – for a total of 8 times today's levels.

This is similar to the result of the brief analysis, in the beginning of this chapter, as to what would happen if all of the projected 9.1 billion people on Earth became industrialised, with emissions levels similar to the First World's today – that was six times current emissions.

But things could be even worse. Developing countries tend to use older, more polluting technologies. And, as oil threatens to run out, more and more power plants all over the world – but particularly in the developing world, where they're being built much faster – are fuelled by coal. This is already causing China's emissions to outstrip even its rapid economic growth.

China emits twice as much greenhouse gas pollution per dollar of GDP as the US, with India at nearly the same ratio. This comparison is made purchasing power parity – basically, internal prices rather than international trade prices. True purchasing prices – which do have some applicability, given the dependence of growing countries on exports – make the comparison about twice as bad again.

So if first China, then India grow to be as rich as America per person, while remaining as polluting per dollar of output as today, their emissions will be roughly 15 times those of the US, or nearly four times the current emissions of the entire world. This would completely swamp any cuts made by others.

China is making notable efforts, having made their stimulus package for the recent credit crunch quite "green", and passing the US in solar cell production. There's a project on the books – but recently delayed for several years – to build a zero-carbon city for half a million people, Dongtan, outside Shanghai.

As with the US 100 years ago, when it was starting to move into world leadership, everything China does is big. But current efforts only lay the groundwork for possible overall improvements in decades to come, even as a new coal-fired power plant or two goes up every week.

China recently announced it will increase coal production by 30% from 2009 to 2015. This implies a doubling of emissions by 2025 alone.

Also, note that the rate of increase in emissions worldwide has itself increased. It was about 1% throughout the 1990s and has been 3% through the first years – almost a decade – of the 21st Century. This could imply that the rate of increase is on track to increase further – to 4% (doubling every 18 years) or 5% (doubling every 14 years). We'll stick with the 3% figure, but keep in mind that there's just as much reason for emissions to get worse as to get better.

The only good news is that 2050 represents the end of the period in which current trends would include both a fast-growing population and a fast transition of people to high-emissions lifestyles. That means the rate of growth of emissions should eventually begin to slowly – the key word is slowly – decline.

By the middle of this century, if economic growth can be maintained, the vast majority of people worldwide will have achieved a developed-world lifestyle. At that point, the growth rate of emissions should level off further, with emissions gradually flattening.

Why emissions won't easily drop

Bringing down global emissions is usually discussed as if it were relatively easy – "just" a matter of will, focus and persistence. Advocates of action on climate change often describe it as needing just a single percentage point or so of economic effort to make a dramatic difference. But it's actually somewhere between the hardest thing humanity has ever done and nearly impossible.

Why? Let's take a quick look at each area – and at the big picture.

For electricity generation (33% of US emissions), only fossil fuel and nuclear plants provide 24/7 "baseload" power and are easy to build, cheap and scalable. As oil runs low, it's largely being replaced by much dirtier coal, which can be liquefied easily where needed (in a highly CO_2-intensive process). Electricity needs are increasing sharply as China, India and others industrialise, and as air conditioning use skyrockets, propelled further by warming.

Transport (28% of US emissions) is largely powered by fossil fuels. When oil runs low, liquefied coal (shudder) is the logical replacement. While trains are less polluting than private cars, they don't go all the way to the end of most trips – and ships and jets, proliferating along with world trade, require fossil fuels.

Industry (20% of US emissions) generates power in a way that resembles electricity generation, but is less centralised and therefore harder to change by fiat. Rapid, forced change in industry is most likely, in the first instance, to cause new and interesting ways for businesses to go out of, well, business.

Agriculture (8% of US emissions) is multiplying in its impact as the population rockets upward and as developing countries "move up the food chain" to consume more meat and dairy, which greatly increases the impact of each such consumer, including through livestock emissions and deforestation. Cutting emissions from agriculture would require cutting global meat consumption in a world that wants more and more meat.

Commercial and residential buildings (6% and 5%, respectively, of US emissions) need changes that are, literally in the case of residences, quite "close to home" and emotional. Imposing construction or retrofitting costs on business, bulldozing old housing stock for newer, more efficient buildings and changing our daily habits are all fraught, difficult processes. Developing countries are throwing up new buildings with amazing speed – and, again, using tried and trusted, high-emitting approaches, not waiting for "green" technologies to become mainstream.

In all of these areas, major infrastructural changes will take decades – with emissions reductions not even beginning until a) replacement approaches are completely ready, tested and in fairly large-scale use and b) regulations, incentives and penalties are in place to make them greatly preferred to higher-emitting alternatives. Every day that goes by before these things happen leads to more infrastructure going in that will be emitting greenhouse gases for decades to come.

If we do tackle these problems, cost is a big issue – and highly controversial. The best current estimate of costs is 1-2% of GDP, as stated in the Stern Review, a figure comparable to military expenditures in most European countries. This estimate can be attacked as too low or too high.

One leading economist who studies climate change, Geoffrey Heal, estimates that producing 25% of US electricity – less than 10% of all US energy usage – from renewables would cost $2 trillion in infrastructure investment, indicating the Stern figure may be low. A great deal depends on who ends up paying the costs.

The Congressional Budget Office has estimated that the Markey-Waxman climate change bill would cut emissions at a cost of just pennies per week for the average household, with lower income households actually coming out ahead.

The developing world has its own considerations. A "green" infrastructure should cost less overall, but any additional up-front expense or delays will seem intolerable to poor countries in a hurry to develop. This is the challenge that is not currently being discussed, let alone met.

Why CO_2 levels won't drop at all

One of the greatly confusing factors in understanding climate change is the difference between various greenhouse gases, and between greenhouse gas emissions as opposed to greenhouse gas concentrations in the air and water.

Here's a quick look at the major greenhouse gases and other emissions that affect climate change:

- CO_2 is the most important greenhouse gas for understanding climate change. It stays in the atmosphere for a very long time; perhaps a third of it may stay in the atmosphere for more than a century. Also, the processes that remove it from the atmosphere and the seas are slowing; and the warming it causes lasts for many centuries.
- Methane (CH_4) is the next most important greenhouse gas. It stays in the atmosphere for an average of roughly a dozen years, then decays to CO_2, which hangs about for many more years, and water (H_2O).
- Water vapour is an important greenhouse gas; sceptics delight in pointing out that minor changes in the behaviour of water vapour as clouds, fog and haze, could swamp the effects of other greenhouse gases. But the only change projected is the retention of more water vapour in warmer air, increasing warming.
- Soot helps trap heat, increasing warming, and also helps melt any ice or snow it falls on. It washes out of the atmosphere with sufficient precipitation but can build up on snow or ice surfaces over years, accelerating their melting.

The problem is that most emissions, and the warming they cause, stay in the environment for a long time. Various recent simulations seem to indicate that emissions and the warming they cause can be regarded as more or less permanent for the next 1000 years.

The situation has been described as a nearly full and slow-draining bathtub with the tap left wide open; the bathtub can easily overflow if the taps aren't shut all the way, or very nearly so.

For the processes that normally absorb CO_2 to have time to work, we would have to cut emissions to one-fifth their current global level per person and stop deforestation, then wait centuries for the current excess to clear. With those processes increasingly impaired, the wait would be even longer – yet there's no prospect in sight of global emissions doing anything but rising sharply throughout this century, and of tipping carbon sinks adding a roughly equivalent impact.

S curves and tipping points

The underlying trends driving emissions upward will eventually slow. The number of people worldwide is set to level off, which means the number of people yet to get rich is several times today's level but not limitless; and growth in emissions per person from those who are already rich is already slowing or, in Japan and Europe, stopping. This means we're looking at an S curve for greenhouse gas emissions rather than just an exponential curve, increasing without end.

S curves, in the abstract, are fascinating. They describe what happens when a phenomenon reaches exponential growth, accelerates – then encounters limits and gradually begins to slow, finally levelling off.

Successful new technologies often follow an S curve for adoption. That's because they accelerate due to demand by early adopters, but face a limit to growth: 100 per cent penetration, as it's called in marketing, of the target population.

So successful technologies tend to have a slow period of growth in which they are purchased, in their early – usually clumsy and expensive – form by early adopters. Some technologies stall or die at this point. (Tivo, anyone? Microsoft WebTV?)

Successful technologies, though, get their rough edges sanded off – they become both better and cheaper. Think of the evolution of the personal computer, from the Apple II to the IBM PC to the mass adoption of Windows PCs and Macs that we see today. An adoption that has been overlapped by the rise and rise of mobile phones, from the clunky, brick-like car phones of the early 1990s to the small, powerful, multifunctional devices available now.

The point in an S curve where a technology or product starts to grow explosively – where it goes from an oddity owned by a few to a normal part of life for millions of people – is called the tipping point. This term is used to describe the crucial point in the growth of a wide range of phenomena. The tipping point is normally reached when a technology or product has penetrated one-sixth of its ultimate potential market – when it has penetrated to within one standard deviation from the mean, to put it in statistical terms.

So think of the high-emissions lifestyle, based primarily on the automobile, as a product. The high-emissions lifestyle reached its archetypal expression in the newly built American suburbs of the 1950s and early 1960s, when a newly built nationwide network of superhighways was not yet overfilled by the giant cars that cruised down the new roads, powered by cheap American and pre-OPEC Arab gasoline. Think of a small-boat-sized "ragtop" – convertible – pulling up to a drive-in restaurant, AM radio tinnily blaring early Beatles tunes, and a carhop – waitress – wheeling out on roller skates to take your order. Of a burger, fries and a milkshake or Coke, most likely.

This fantasy – which was also a reality, one that I remember from my childhood in Southern California – didn't last long in its pure form. Americans themselves rebelled against it quickly enough, from the counter-culture revolution and civil rights movements of the 1960s to the environmental movement of the 1970s. Americans then tried to return to the same fantasy in the conservative ascendancy of the 1980s and beyond. ("Greed is good", quoth Michael Douglas, playing financier Gordon Gecko, in the 1980s film *Wall Street*.) The Australians cheerfully adopted all of it, with their own twists. Europeans always looked at the high-emissions lifestyle from one remove, while taking on much of it, and the Japanese didn't have room on their islands for the unadulterated version.

Yet the high-emissions lifestyle retains its attractive power. You can see it in the growth of trends associated with its more positive and more negative sides – automobile ownership, denoting wealth and freedom, reaching one car per adult in America; obesity, the result of too many burgers, fries and Cokes, becoming ubiquitous in the US and Australia. It seems as if more than a billion people each in China and India are all trying to get their own convertible and roll down to their own drive-through – probably while chatting on mobile phones and with iPods plugged into the car stereo, of course.

So the high-emissions lifestyle is an incredibly successful and seductive product. Regimes have been overturned (the USSR) or transformed (China, India) by people eager to pursue it. The worldwide explosion in emissions is simply the result of people moving to it as fast as humanly possible.

Ultimately, the S curve of emissions growth we're on will peter out – if nothing interrupts its growth first – when humanity reaches its natural limits of population, which seems to be the 9 billion or so we're projected to reach in 2050, and when the vast majority of humanity is living the high-emissions lifestyle, which would, on current trends, occur later in the century.

The key marker, and key statistic of the high-emissions lifestyle is car ownership. As a rule of thumb, when a country has reached the point that one-sixth of households own a car, it has tipped toward the high-emissions lifestyle. If circumstances allow, within a few decades, five-sixths or more of its households will own a car, and most of the other accoutrements of the high-emissions lifestyle as well. (A large house far from work, so you have to drive to it; various gadgets; and a not particularly healthy diet with lots of meat and processed foods that have travelled long supply chains to reach the consumer.)

The US is super-saturated with cars. It passed the tipping point of one-sixth of households owning cars in the 1930s, despite the Great Depression; passed the point where five-sixths of households own a car in the 1950s; and now there are as many cars and light trucks registered in the US as there are adults. Americans live in huge houses, by world standards – the average is about 2000 square feet (200 square metres), twice the European norm – drive many miles to work and school, are obese and becoming more so at a world-leading rate, and so on.

Of course, the high-emissions lifestyle is marked most notably by high emissions, with Americans, as previously mentioned, emitting 20 tons of greenhouse gases per person a year – double the next highest large grouping, Europeans. This is clearly unsustainable.

Yet even in Europe, with much more environmental awareness and concern than the US, officials are finding emissions reduction efforts stymied by the steadfast refusal of drivers to drive less. This despite fuel costs triple the US level, high vehicle taxes and insurance costs, and much more constrained road networks in most countries.

There is no easy way to stop the move to the high-emissions lifestyle. It is almost certain to continue to expand through global populations until it reaches limits of one kind or another.

A more realistic curve for emissions

So what should the IPCC Scenario curves really look like, if we include the well-established global move to Western lifestyles?

In order to come up with a curve that's impossible to call pessimistic, while still giving a flavour for the true direction of emissions, I've used the following assumptions:
- Populations grow to 9.1 billion people by 2050 – and then start to gently decline, as projected (without taking climate change-caused problems into account in either direction);
- Economic growth and rising emissions continue until all 9.1 billion people are living a European lifestyle, with emissions at current European levels – which American emissions today

exceed by a factor of two, and which Chinese and Indian emissions are on track to exceed by two to three times that again;
- Emissions per person then stay level, and total emissions track population growth.

Figure 4-4 shows the result: basically, an S curve with emissions peaking at nearly three times today's levels. Though certainty as to specific temperature effects decreases as greenhouse gas levels in the atmosphere increase, this curve almost certainly represents more than 7°C of total warming, plus strongly destructive acidification of the oceans. This is even before tipping of carbon sinks, as described in the next chapter, is considered.

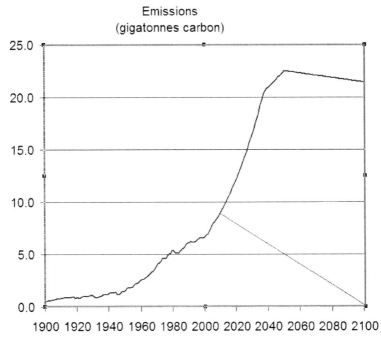

Figure 4-4. Greenhouse gas emissions with emissions per person worldwide converging to European levels by 2050; straight line shows near-50% cut by 2050

Also shown in Figure 4-4, for contrast, is the cut in emissions – a drop nearly in half by 2050, and a reduction of 100% by 2100 – estimated to be sufficient to keep warming beneath 2°C, the most aggressive goal currently considered realistic. (Cuts that are themselves insufficient and ineffective in preventing runaway climate change.)

The big difficulty in any such exercise is accounting properly for what the developing world will do, since emissions per dollar of output in China and India, as the largest examples, are so much higher than those of the high-emitting US. They are thereby on track to completely swamp current global totals in the coming decades.

It's easy to say that "of course" the level of emissions per dollar will drop. It almost certainly will – but only with great difficulty to even current Western levels. There's no way for emissions to drop in the short term if there's any growth at all in the developing world, let alone the six to 10% annual growth expected for many years to come.

The higher line on this graph reflects one possible scenario in reconciling several contradictory trends, which can't all continue at once:

- Continued growth in emissions of 3%/year until emissions converge; emissions may actually increase faster during the most intense years of growth in the developing world;
- A convergence of average world emissions at current European levels of about 10 tons/person/year; the US and Europe are planning cuts, but much of the (far larger) developing world is on track to far exceed even American levels of 20 tons/person/year;
- World population peaking at 9.1 billion people in 2050, which is the mid-range of admittedly uncertain projections; climate change and other trends, many described in this book, could lead to far different outcomes, many of which could feature both higher birthrates (as economies fail to mature) and higher death rates, with the total population volatile.

If the curve were drawn so as to converge to *American* levels of emissions per person, which would still require strong cuts in the current trend of the developing world, it would rise faster, and for longer, to a level nearly twice as high as the curve shown – and only then begin a slight decline.

Also, many of the emissions reductions discussed today shift the peak of the curve out by perhaps a few decades, but don't reduce the height of the peak much, if at all. Any scenario in which world economic development continues is accompanied by several times as much farm and grazing acreage, several times as many airports, several times as many cars and miles of roads, several times as many office buildings and several times as many 1000 to 2000 square foot (100 to 200 square metre) homes, replacing much smaller homes, as we have today. In the scenario reflected in the graph, these numbers will be about three times what they are today.

This by itself explains why pressure for development is so incredibly strong today. The world needs many times the infrastructure it has today to accommodate people who are on the way to reach a rich country's lifestyle by the middle of this century.

In this projection, which is actually below a "business as usual" projection, people in 2050 will be emitting roughly three times the greenhouse gas emissions we do now. The world's forests will be almost gone, done for by direct deforestation and by damage inflicted by climate change. Desertification and lack of fresh water will be cutting crop yields worldwide. The seas will have ceased supporting anything like the richness and variety of life they still manage to host today.

If you tell people that this is unsustainable – that the world would do well to increase infrastructure by, say, 25% before we run out of room, water, years without climate change-caused natural disasters and so on – they don't all stop and say "Oh, OK, we best not push the limits". They seem to say, at least to themselves to themselves, "If the drawbridge is about to be pulled up, we better hurry and get inside the castle" – into the rich world – "fast." (The same kind of "tragedy of the commons" reaction that has depleted marine life, forests, fresh water supplies and more.)

IMechE and others sound off

I've not yet seen other projections that take into account both up-to-date emissions trends and likely tipping of carbon sinks, as in this book – though this is something the IPCC should be providing. However, some important and reputable organisations are starting to "take matters into their own hands" and make predictions about emissions and climate change independent of the weak IPCC Scenario-based projections. The Institution of Mechanical Engineers in the UK, known as IMechE, is one early and important example.

This group represents a profession, mechanical engineering, that builds a great deal of infrastructure, including infrastructure that directly or indirectly lasts for hundreds of years. (The first line on what is now London Underground went into service nearly a hundred and fifty years ago, and Heathrow Airport first came into use in World War I, for instance.) This gives them a perspective well beyond the somewhat artificial deadlines, such as 2050 and 2100, that most other commentators impose on evaluating the consequences of climate change. (Present company included.)

In a report published in February 2009, <u>Climate Change: Adapting to the Inevitable</u>, IMechE took a fresh look at climate change, with three key assumptions:
1. Efforts to reduce emissions, even among countries that made commitments under the Kyoto Protocol, have failed so far, and are likely to largely fail going forward;
2. Fossil fuels – coal, oil and natural gas – are so familiar, cheap and convenient that most reserves will be used, emitting their carbon into the atmosphere;
3. While the specific timing can be disputed, these factors alone will lead to a temperature increase of roughly 5°C, and accompanying sea level rises of several metres, during the lifetime of much existing infrastructure and many future IMechE projects.

IMechE commissioned independent assessments of the direct effect of emissions to arrive at these conclusions, but they used the IPCC's lowball projections and ignored the effect of feedbacks from nature. Neither did they address many of the impacts on nature and on human societies of the rise they projected, let alone the further rises from carbon sinks tipping.

Even so, their conclusions were quite dramatic, and in rough agreement with the comparable part of the more comprehensive analysis presented here. The main difference from the 2007 Consensus is that IMechE assumes that humanity won't significantly slow, let alone stop or reverse, emissions, and that the world will soon see temperatures many degrees higher and sea levels many tens of metres higher.

You can download the report for free at the Institution of Mechanical Engineers Web site; see Appendix C, Selected Resources, for the link.

For an established, reputable professional group like IMechE to present conclusions like this has repercussions going forward. (More so, perhaps, than the conclusions of an independently written book like this one.) Mechanical engineers and other professionals will be forced to deal with the IMechE's conclusions in their own planning, or face possible legal and reputational consequences for having ignored these inherently credible conclusions.

IMechE is not the only professional voice to be raised. *The Lancet*, one of the world's leading medical journals, undertook a study of the public health effects of climate change in May 2009. The study found that temperature rises "are likely to exceed the safe threshold 2°C above pre-industrial average temperature" – as also asserted, though more strongly, throughout this book. It calls climate change the biggest threat to human health in this century, "a clear and present danger" that will affect billions of lives.

Climate scientists are also stepping outside the too-conservative IPCC projections. The June, 2009 issue American Meteorological Society's *Journal of Climate* – a leading, high-profile, peer-reviewed journal – published a study from the Massachusetts Institute of Technology predicting warming of 5-7°C by the end of this century, similar to my own analysis in this chapter. This is from greenhouse gas emissions and direct deforestation alone, without taking into account runaway climate change, whose likely effects are described in Chapter 7.

Even governments are getting involved. The US Government's Global Change Research Program recently released a landmark report, *Global Climate Change Impacts in the United States*. This report was delayed for years by the Bush administration and finally released by the Obama administration. It points to higher temperatures, stressed water resources, interaction with other social and environmental trends and the crossing of thresholds that will change ecosystems and regional climates. One particularly compelling comparison: even under its relatively conservative projections, Illinois in 2100 will be as hot as Texas is today.

The British government announced new figures from the UK Climate Impact Programme projecting a 6°C temperature increase in the south of the country by 2080 and London peak summertime temperatures of 41°C (105°F), 10°C more than today. Greater drought and flooding are both expected as well. Nearly a million homes are projected to be at risk of flooding *by 2035*, with much worse impacts in subsequent decades.

All of these prognoses are still based largely on the 2007 IPCC Report, which does not recognize much glacial melting nor tipping of climate sinks. Their results, though dramatic, are simply straight-line projections of current emissions trends, which run ahead of any and all of the IPCC's 2007 projections. (Which means, of course, that the projections were awful.)

So it's among the most promising signs that Chris Field – one of the IPCC scientists invited to receive the Nobel Peace Prize in Oslo – recently stated that "the actual trajectory of human emissions" are "now outside the entire envelope of possibilities that were considered" in the 2007 IPCC Reports, which of course prominently featured projections based on the 2000 Scenarios. This shows that at least one of the thousands of IPCC scientists, a leader in the IPCC, really "gets" at least some of the argument expressed in this chapter.

But a great deal of damage – perhaps a decisive amount of damage – has been done. "You only get one chance to make a first impression", the old saying goes, and the IPCC 2000 Scenario-based projections have been imprinted on the minds of an entire generation; people who saw the resulting graphs as part of their first awareness of the (greatly understated) seriousness of climate change. They'll hang around in books, textbooks and on Web sites for years to come.

And the issue of whether climate change might already be runaway, or even quantifying what this deadly serious shift would entail, is not yet being seriously addressed – leaving a hole one could drive a truck, or at least this book, through.

While this book and other voices may be a minor beginning, it will take a new IPCC Report, using much more realistic numbers, to begin to reverse the earlier misimpressions globally. The next Report is now due in 2014, so that's the first opportunity to finally start to get this right.

Can emissions really keep growing?

There are three large sets of unknowns that could prevent the scenario represented in Figure 4-4 from being fully "achieved" – remembering that the projected emissions are a severely unwelcome side effect of otherwise desirable economic growth.

Before beginning to describe them, though, it's important to point out that it's the next few decades that are crucial. There's a saying in politics that's summed up in the acronym EMILY: Early Money Is Like Yeast. Early donations attract more money and cause the whole campaign to take off. The same is true for emissions: early emissions have a "chain reaction" effect, causing immediate damage and contributing to greater future emissions and faster and more complete collapse of carbon sinks. All of the issues mentioned below tend to, at best, slow the growth of emissions in the near term, with any actual decreases only taking place some decades down the road.

Having said that, the most welcome constraint on the scenario represented by Figure 4-4 would be a "good" demand-side issue: reduced demand due to an intense surge of effort all around the world to cut emissions.

This is possible, but it hasn't happened yet, and would demand two things that are unlikely. The first is discipline in, and true leadership by, the developed world. Some of this is emerging, but so far not nearly enough.

The second is a willingness in the developing world to tolerate slower growth while truly "green" growth can be defined, implemented and deployed on a global scale. Only the most intense joint commitment in this regard would meaningfully affect the crucial next one to two decades.

Even with emissions thus lowered, massive geo-engineering efforts would still be needed to remove current greenhouse gas pollution from the environment and reverse the path we're on.

The next, less promising possibility for constraints on emissions is a supply-side problem – which some see as a good thing, but which would actually be bad both for people and the environment. It's been suspected for several years now that oil production may now be in ongoing decline – called Peak Oil. However, the easy, if awful, answer to that is to liquefy coal.

Some say that coal reserves are also not what they're believed to be, and that coal could run out in the next few decades as well. But the end of both oil and coal would be a cure (for greenhouse gas emissions) that rivals the disease in its ill effects.

The last years of either oil or coal, let alone both at once, would be a disaster. The dirtiest possible deposits, along with peat – even more polluting than coal – would be dug up and burned as energy prices skyrocketed. Unless alternative energy could be brought online at huge scale, and very rapidly, a truly epic worldwide depression would ensue, only worsened by accumulating and accelerating damage from climate change.

A possible end to oil and, more remotely, coal supplies is not an argument against urgent action on efficiency and alternative energy sources to combat climate change – it's the strongest possible argument in favour. Simply trading off between one global disaster and another gets humanity nowhere.

Along with supply side issues and the possibility of self-discipline on the demand side, there is another potential demand-side limitation. This is a series of disasters – environmental, in food production, fires etc. – so large as to severely reduce human populations and/or growth in demand. However, most of the potential disasters will hurt the poorest the most – those who emit the least. So even extensive disasters would do little to reduce total human emissions until later in the century, after the course of the environment is for the most part determined, rather than in the nearer term.

Chapter 5. Emissions Growth Worldwide

> In this chapter
> - The UK and Europe
> - Emissions and the US
> - China and the developing world
> - Russia and Canada

To understand where emissions are coming from and how we might begin to fight them, we need to look more closely at the largest emitters in the world, both past (the US and Europe) and future (the US and Europe plus China and India).

Before looking at how emissions are expected to grow from different regions, though, let's review the good news of which the bad news is a side effect.

Our current problems are actually the result of something that should be wonderful – and in many ways is. The Industrial Revolution has made it possible for even many of the worse off people in an industrialised, or "rich", country to live in a decent sized house, have a car, TVs, a PC and enough food to be comfortable. Children can be clothed and schooled, and everyone can get decent medical care, if the society is managed fairly. At the same time, the wealthy in industrialised countries are almost unimaginably well off by past standards.

Not only rich countries have benefitted. A great, little-told story of our time is that the number of very poor people worldwide has been dropping steadily over the last several decades. At recent rates of improvement, extreme poverty – people living on less than a dollar a day is the usual measure – could be just about wiped out in a few more decades. (Though with the upcoming problems described in this book, things will almost certainly get worse rather than better.)

Trying to force people away from the move to the rich countries' high-emissions lifestyle is cruel, unfair and fruitless. People will give up almost anything to secure a higher standard of living for themselves and their children.

The move to at least a modified version of the high-emissions lifestyle is partly a human rights issue – relating not only to the "pursuit of happiness" at the high end but also to the avoidance of absolute poverty at the low end. Only rich, high-emissions countries can afford a decent standard of living even for their poor and disabled.

We also need to recognise that the growth that has benefitted so many was spurred by cheap energy – coal, oil and natural gas just sitting there in near-surface deposits, waiting to be dug up or drilled for and burnt. This is such a valuable resource that today's advanced economies have used up most of the easy to get to oil, in particular, on the whole planet in order to, literally, fuel their growth.

The trick, if it's humanly possible, is to have a high-emissions lifestyle without the high emissions. It is indeed possible, using current technologies, such as existing solar power facilities and wind turbines, and reasonable extensions of them, such as very extensive smart grids. All of this has to be deployed on a previously unheard-of scale.

The real concern is the related issues of the speed of the transition and its cost. The faster the transition, the greater the cost, especially where serviceable but polluting machinery gets taken out of service and replaced by "green" alternatives. This transition will add to energy costs – though how much of that shows up directly on consumers' bills is an open question. How to preserve growth in a period of more expensive energy and more expensive infrastructure, especially in developing countries, is a huge challenge.

The UK and Europe

To understand today's emissions trends, we should begin by looking at the UK and Europe, because the UK is where the Industrial Revolution – the source of much of the world's current wealth, as well as of our current crisis – began. The UK and Europe are also where at least some progress in cutting emissions is being made, with successes and failures that the rest of the world can learn from.

European countries have found a better balance between an advanced economy and emissions than the US. The typical Brit or continental European emits about 10 tons of greenhouse gases per year, half the level of the typical American. One way this has been done is through very high fuel taxes; petrol typically costs three to four times as much in the UK, for instance, as in the US. Homes are smaller and air conditioning use is much less, even where seasonal temperatures are similar.

As a set of advanced economies with a roughly static population, western European countries tend to add greenhouse gas emissions at roughly half the rate of economic growth – one percent of economic growth results in one half a percent growth in greenhouse gas emissions.

However, with climate change looming, and within the Kyoto framework, most European countries have agreed to a formula called 20/20/20 – a 20 percent reduction in emissions from 1990 levels and 20% of energy from renewable sources by 2020. And they've made remarkable progress toward it, cutting greenhouse gas emissions by several percentage points while growing their economies by a few percentage points a year.

European progress under the Kyoto accords does include a trick along with the treat. The accords count emissions cuts from 1990 because Eastern European emissions slumped just after this point, with the fall of communism. Taking credit for these accidental reductions gives all of Europe – as well as Russia – a leg up.

And it's easy to criticise the overall effort, as well as missteps along the way; early and strict European commitments on biofuel use have caused a great deal of rain forest acreage to be cut down to grow biofuel crops, for instance, a net environmental loss. But still, Europe has actually managed to grow economically while shrinking its emissions footprint, even in the face of economic competitors not bothering to do the same.

How has this been done? Germany may be the best example. To meet strict regulations, Germany has become the world's leading buyer and user of solar panels – despite being far from ideally positioned to use solar power. And Germany is refitting millions of older homes across the country with insulation in a 20-year programme. Germany also leads, alongside Denmark, in wind power. It has recently been the world's leading investor in renewables.

With current economic difficulties, some European leaders are arguing for easing of the 20/20/20 goal; some countries make supportive noises in favour of the goal but look like they'll fall short.

But there's still, at this writing, everything to play for, as the Brits say. Going it alone – without the benefit of other parts of the world operating under the same constraints – has made things far more difficult. What the Europeans have accomplished already, and may yet accomplish in the coming decade, is remarkable.

Emissions and the US

In contrast to Europe, the US has so far significantly undermined its position of world leadership by inaction on climate change.

This was not a failure of the Bush administration alone; at the turn of the century, US public and political opinion were strongly against any meaningful involvement in a Kyoto-style process.

At a basic level, this was, while immoral, entirely in line with the short-term interests of the US, with American emissions per person immense – double Europe and Japan's, four times China's, and so on. (The real picture is even worse, as the US, consuming a big chunk of China's output in particular, indirectly owns a big piece of their emissions as well, plus the shipping and air freight to transport it.)

A logical position for a neutral in climate change discussions to support would be for the US to cut its emissions in half by, say, 2020 – just about impossible though that would be – at which point a sensible conversation with the rest of the world could start.

The US legal system also pioneered the "polluter pays" principle for pollution damages and cleanup. In US courts, it doesn't matter who knew what when; if you caused the damage, you pay for it. (Actually, prior knowledge matters in a different way; it potentially makes you liable for huge damages over and above costs.)

The US is responsible for about a third of historic greenhouse gas emissions. It knew more, earlier, at a higher level about climate change damage than anyone; a crack scientific team called the Jasons foresaw it all in a secret report delivered to President Jimmy Carter in the late 1970s. Future opening up of Arctic sea routes, as is happening today, also became a topic of discussion in Western military circles at that time.

But neither the scientists nor the generals were first. Insurance giant Munich Re called attention to climate change in a 1973 publication, citing "the pollution of the Earth's atmosphere" by CO_2 and pointing out that the impact of climate change "on the long-range risk-trend has hardly been examined to date". Perhaps this alarm helped get the Jasons focussed on the problem.

The conclusions by the Jasons were supported by a team of American climatologists in the early 1980s, by James Hansen of NASA in the 1980s and since, by the US Navy's North Pole sea ice data in the 1990s and on and on. The US government and businesses also have the most advanced technology – to reduce and clean up emissions – and the deepest pockets of anyone in the world to pay costs and damages from.

But the US sees China as an economic and strategic rival today and as a potential enemy tomorrow. (With Taiwan potentially set to become the current century's Berlin.) So, in addition to people's typical reluctance to change or to incur expense, the US has been very loathe to give up anything today that didn't also slow the Chinese down, as would have happened – entirely reasonably, from a neutral point of view – with Kyoto.

So the damage already done and due to be caused by American emissions, the long-time refusal of the US to get involved in Kyoto, America's continued slow response to the need for change, and its fixation on technical fixes make short-term practical sense. It even makes a chilling kind of short-term strategic sense: China is set to be devastated by a series of environmental catastrophes, with climate change the major, but far from the only problem, in the next few decades, while the US could probably ride out the first couple degrees C of warming with far less damage.

The recent trend in the US, unconstrained as it has been by any international agreements, is for emissions to increase forty per cent by 2050. This sounds like a lot, but the size of the economy should double at least once in that time, and population growth of about 20% is expected. The forty percent increase reflects a much lower increase in emissions per point of economic growth than the current roughly one for one relationship pertaining in the developing world.

With much better understanding of climate change's effects (thanks to Al Gore in particular), and with President Obama and a more strongly Democratic Congress now in office, there's now far more willingness to tackle climate change. Obama has rather cleverly tied fixing climate change to energy independence, making them together one of his top three goals for the US, along with economic stabilisation and health care. Energy independence and acting against climate change are ends that share almost all the same means.

It remains to be seen how much President Obama and Congress will act on the domestic priority for energy independence, or exactly where the US will land on climate change in international discussions. One is reminded of the Winston Churchill quote in which the US can always be counted on to do the right thing – but only after it does everything else first.

China and the developing world

The role of China is important on its own – with China having recently become the world's leading greenhouse gas emitter while remaining the fastest-growing – and as a proxy for all developing countries.

There are billions of people in the world wanting to take on a Western lifestyle and more or less on the way to doing it. China is the (very) large country that's actually well launched on the way – while racing, perhaps purposefully, against a host of environmental and demographic challenges.

As the last large Communist country, and as the first major country set to "get old before it gets rich" – that is, due to have a huge percentage of its population above retirement age, but without the national wealth to more easily support them – China has problems that probably outweigh climate change in the minds of its leaders. Whether they should do so is another matter.

China is also important because its position in relation to the challenge posed by climate change has been so poorly understood, and because remedies currently proposed for climate change fit China so poorly.

Developed countries, which all solutions proposed so far have been addressed to, have already developed their high-emissions lifestyles. They're able to have a few percentage points a year of economic growth with only about half of that amount in emissions growth. With hard work emissions may even be able to be cut while the economy continues to grow, as the Europeans have demonstrated.

For China, though, its six to ten percentage points a year of economic growth comes with a roughly identical increase in emissions. (Six percent growth is a minimum for employing young people entering the workforce or moving into towns and cities from the country, so it's the bare minimum of growth needed to support political stability.) In fact, with coal making up more and more of China's energy mix, emissions are increasing at an even faster rate than growth.

Emissions rise in lockstep with growth because, as a developing country, China's growth comes from clearing land; building new homes, offices and factories; producing more meat; people getting their first-ever cars, televisions, PCs, central heating and air conditioning; or taking their first trips overseas (or newly beginning to fly overseas ten or twenty times a year on business) – all bound to increase emissions rapidly.

China is building infrastructure with all this in mind. It's pushing hard to get its share of the world's oil output and other vital commodities, trying to overcome its Johnny-come-lately status. But what China does have is coal – perhaps a couple of centuries' worth. And they're using it, building an impressive – and truly frightening – one to two new coal-fired power plants a week, with no provision for carbon capture and storage. (Which may never prove cost-effective, at least within China's needs for fast growth and low cost, in any event.)

China is even higher-emitting than might otherwise be the case for two additional reasons. You don't grow an economy six to 10 percent a year – doubling its size every 10 years on average – by re-inventing the wheel every time you plan a town or a factory. You use existing, proven – and, unfortunately, highly emitting – approaches. And you don't over-regulate, even if that means higher future emissions from your infrastructure. So the speed of China's rise makes reductions nearly impossible given today's technology.

China has also been growing largely by manufacturing for export. China is separated from its customers by barriers of distance (meaning time and cost), language and custom. The only way for China to get orders is to be the low-cost provider by a big margin. The very low bids that Chinese suppliers can make are known as the "China price". This means cheap power, cheap factories and cheap processes, as well as speed and flexibility. (Whole factories are put up or knocked down as the order book changes.)

None of this fits with emissions reduction, which occupies brainpower that's needed elsewhere, adds cost and complexity and, just as importantly, takes precious time.

Viewed this way, the Chinese standpoint on climate change makes sense. It could be paraphrased as: if you give us (not sell us) lower-polluting technology ready to deploy, and finance its installation, we'll consider using it. But don't expect us to divert resources from growth to solve a problem that you in the West caused, and continue to exacerbate.

There's no way to stop China from growing, and no easy way, in the timeframes involved, to support them in growing in a significantly less environmentally damaging way.

A final major problem with dealing with China's emissions is now resolved, but has already had some lasting effects. Until the last couple of years, China's emissions were underestimated by a factor of about two. It was only the Kyoto Protocol's requirement that China monitor and report its emissions – criticised by conservatives in the US and elsewhere as meaningless – that caused the true numbers to emerge. It's even been said that China is deliberately running up its emissions now so as to have a higher starting point from which to negotiate slower growth in the future, followed eventually by cuts.

What this means is not only that the 2007 Consensus had a truly crucial fact, China's emissions, hugely wrong. It also means, among other things, that the action plans that came out of that consensus all required relatively modest overall changes, as appropriate to a world in which slower-growing Western economies were the driving force behind greenhouse gas emissions. The difference with countries that were growing economically and in emissions by nearly double-digit figures was not thought through, and consideration of the much more wrenching changes required to reduce or stop emissions growth from developing countries was set aside.

The belief has been that the West could go first, take ten or twenty years to develop lower-emitting technologies, then share them with –

or sell them to – developing countries, thereby capping everyone's emissions growth. (Never mind all the land clearing, livestock raising, new car buying, flying and so on that would continue to grow in developing countries, generating huge emissions.)

The new numbers show that this is not on option. By midcentury, China's economic output – and emissions per person – should double twice, to US levels. The population should double once, all with accompanying emissions growth. That will make Chinese emissions alone equal to (at European emissions levels) or twice (at American emissions levels) today's world total, rendering any emissions cuts the West could make moot.

Even worse, China's emissions per dollar of production are currently about twice the West's. So China is actually on track to reach more than eight times US emissions when it reaches equivalent gross domestic product, depending on its peak population. This trend will have to be brought into check as well – and means the world as a whole is very far indeed from actually reducing emissions.

China is making immense efforts on areas such as wind power and solar power. They are discussing a target of 20% of energy from renewables by 2020 – matching Europe's commitment. However, this is 20% of a vastly increased generation base, with 80% of this larger base coming almost entirely from coal. The new commitment is at a scale to accommodate a share of the country's future growth – not cut sharply into the rapid increases occurring today.

It would take enormous restraint for China to only double emissions once per person, reaching European rather than US levels, and then hold emissions level. Planning decisions and the desires of its own populace are taking China toward an American model instead, but with older and even more highly emitting technology.

Ignoring China wasn't really an option even from the previous, artificially low estimates, and the attempt to do so is hard to comprehend. Perhaps it makes an odd kind of sense as a desperate attempt to maintain optimism about global prospects for improvement, along with a dose of post-colonial guilt.

What needs to be understood is that China's rise to 2050 alone represents a trebling – from one billion, mostly in the US and Europe, to three billion – of the number of people living a Western lifestyle. So the huge and potentially unsustainable environmental costs the world was carrying to support one billion intensive consumers are now trebling as well.

And behind China is India, with even more people and currently at half China's emissions per person – and growing fast, though not quite as fast as China. India faces all the same concerns as China in terms of growing by being a low-cost, and therefore high-emissions, producer. India, lacking China's draconian population control policies, is quite likely to add a fourth and fifth billion-strong cohort to those living a Western lifestyle – and generating Western levels of emissions per person, if not higher – by the middle of the century.

It's also at about mid-century that the Himalayan glaciers may give out – and aquifers that help support Pakistan, India, China and southeast Asia may largely give out well before that. So the food-growing capacity of an area home to half the world's people may plummet by 50% or more, sending world food prices into orbit and threatening hundreds of millions with starvation.

China is trying to get in front of this problem by building dams to serve much of the function that its glaciers currently do. The country recently announced plans for roughly 60 new reservoirs, some of which might have their water storage underground to reduce evaporation. Dams cause their own environmental problems, and the weight of water stored behind a Chinese dam has even been blamed for recent severe earthquakes in the region around it. But even these downsides might be worth it to reduce – not eliminate – the impact of the loss of Himalayan glaciers.

The planning for these dams also seems to send an important signal: China is planning for the Himalayan glaciers to go. This is important; only the US and China, together, could hope to lead the world to stop emissions and reverse climate change impacts to the extent needed to save these glaciers. If China is investing to adapt to their disappearance, that's a big signal that China is not planning on any such effort – and without China, such an effort can't happen.

My comments here are not intended as criticisms of China, India and other developing countries. They are simply following the steps laid out by the West to achieve a better life for their citizens. If anyone should be criticised, it's the West. Leading countries achieved the means to ensure a decent life for their citizens by the 1980s – since which point economic growth and profits growth have slowed. And the OPEC embargo of the early 1970s clearly showed the dire need for new approaches to energy. (A warning re-iterated by the September 11, 2001 attacks, made possible in the final analysis by oil money.)

But the West, in particular the US, "doubled down" on its commitment to fossil fuels and actually cut funding for alternatives, beginning with the otherwise much-admired Reagan administration of the 1980s, and continuing under the three Presidential terms of the (oil-rich) Bush family.

Energy efficiency went out the window – a window which, in so many buildings in the US, and now around the world, is sealed shut so the "aircon" can work better. Developing countries have little choice but to follow the path set by those that are already developed. The West is reaping the whirlwind, literally; not only of its own profligacy, but of the multiplying effects as its model of development is adopted by billions of additional people worldwide.

In addition to China and India, countries on the development path include Indonesia, which may reach half a billion people by 2050; Brazil, at 190 million today and growing fast; Pakistan, with 165 million today; Bangladesh, with 150 million; and smaller countries that contain two billion mostly poor people today and are expected to add nearly a billion more by 2050. If they don't have rapid economic growth, world poverty will not only continue but rise rapidly.

As these additional countries develop, they'll perhaps, among themselves, match the emissions of the world today by 2050 in reaching European emissions levels, as they go through the heavily greenhouse gas emissions-dependent phases of their growth.. If headed to American levels, they'll double that number in the subsequent decades.

But China, in the middle of hockey stick-shaped economic growth and development, and a model for other developing countries, is the burning issue today. By itself, China eliminates any hope that emissions will be reduced, short of a wartime-like global effort or actual world war, between now and mid-century.

Russia and Canada

Just as China presents its own set of problems and also exemplifies those of all developing countries, so Russia presents specific problems and exemplifies the problems of another group of countries, regions and interest groups: those that benefit – some absolutely, some in a relative sense – from climate change.

Just the thought that any country might benefit from climate change is enough to make one angry. The thought that a country might actively foresee the resulting advantages, plan earnestly to exploit them, and even fight – figuratively or literally – to protect them, seems outrageous.

Yet that's exactly what's likely to happen over and over again as climate change develops. In fact, given the scale of problems that may be unleashed in the next few decades, there may be only one word that truly fits these seemingly bad actors: survivors.

Russia has already shown strategic insight and a willingness to act boldly in relation to climate change and opportunities it provides. The melting ice cover at the North Pole produces a unique historic opportunity for nations with northern exposure to claim dominion over vast regions that are likely to contain much precious oil and gas.

And so the Russians have done. To pursue their interests, they have created an Arctic Directorate, established by a much-decorated former KGB officer, polar explorer and head of the KGB's successor, the FSB. (The US Pentagon has a Space Command, but nothing for Earth's own north.) The Russians are establishing new military bases in the region and have warned off NATO and others from "interfering".

In the summer of 2007, as the rest of the world was reacting with amazement to the extent of North Pole sea ice melting, Russia launched a mini-sub mission to the North Pole. The aquanauts planted a tiny flag on the sea bottom and claimed about 50% of the entire North Polar region for the motherland. The claim extends the overall territory claimed by what was already the largest nation on Earth by about 10%.

The newly Russian area will be ice-free in a few decades, open to shipping and exploitation. "Send lawyers, guns and money", the Warren Zevon song says; only Russia was prepared to act decisively, co-ordinating military, scientific and legal efforts, to take advantage of the melting ice.

Their claim is quite plausible under international law, which holds that undersea features such as ridges that extend from a northern country into the polar regions can be considered extensions of a country's sovereignty. Since only the Russians have ever looked in the area in question – and since, arguably, no one else now can transit, let alone do research in, Russia's new waters without her permission – no one can credibly dispute her assertion that the undersea geography supports the claim.

More to the point, the Russians did things in just such a way that the Canadians and Danes, masters of Greenland, will find it easier to make similar claims than to contest the Russian approach.

Canada has, in fact, negotiated a joint claim to the Arctic with Russia and Denmark – leaving only a sliver for America. And Canada is already in contention with the US in the North. Canada claims that newly opening waters between the mainland and nearby Canadian islands are Canadian territorial waters, a seemingly reasonable assertion that the US disputes. (If the shoe was on the other foot, there's no doubt what the US would say.) Canada is actively building up its military and naval presence in the region in rivalry with both America and Russia.

And what, then, about the US? Neither then-President George W. Bush nor Alaska governor Sarah Palin responded to the Russian claim. Wary of international bodies, the US is not even a signatory to the Convention of the Law of the Sea Treaty, the international agreement under which territorial claims in the region are resolved.

What does this tell us about climate change? Simply that it's now in Russia's national interest for it to continue. Also in Canada's, Norway's, Denmark's and, to a lesser extent, America's.

This was already true in any event. Russia in particular has expanses of frozen plains as large as any other country on Earth. Thawed, they could support farming and hundreds of millions of people. Russia would almost have to see them populated as part of defending them; China is jealously eyeing large expanses of eastern Russia, and many Chinese already live and work there.

Like the US, Russia also benefits competitively from climate change in the first couple of degrees of warming, in comparison to China. But with further warming, Russia is far better off than the US, which could end up hungrily eyeing Canada in the same way that China is looking at Siberia. Russia stands to become more habitable and richer even at three and four degrees of warming, and to suffer far less than any other large country even beyond that.

Unfortunately, the tactics needed to fight climate change, on a planet that's already tipped into runaway status, will of necessity be quite radical. They would generate opposition even if all the nations involved had the same underlying interests. As that's far from true, achieving consensus on action may be impossible.

Not only will Russia not help – she may actively oppose efforts to fight climate change. (A few other countries, more afraid of seeming awkward, might be quietly cheering Russia on.) And Russia has both the right and the means to resist.

It has the right to do so because geo-engineering efforts should, in principle, have the consent of everyone affected by them – in a practical sense, at least the larger and more powerful countries or groups of countries. Russia may well not consent.

And they have the means to make it stick. Russia has a huge nuclear arsenal and strong conventional forces when compared to its neighbours, save only China. Russia is unlikely to feel embarrassed to rattle even its nuclear sabre to stop anti-climate change geo-engineering efforts.

So the rest of the world must make it in Russia's interests to be involved, and all concerned will need to be more than a bit lucky in how the situation evolves. If progress is stopped by a missile-waving contest between Russia and the US, the world's chances will be between slim and none.

Chapter 6. How Carbon Sinks Have Already Tipped

> In this chapter
> - Deforestation and "thinning" of greenery
> - Carbon sinks defined
> - Losing the Arctic "ice cap"
> - The yedoma erupts
> - Damage to other carbon sinks

One of the big worries about global warming – the underlying process that leads to climate change – is that it will become a runaway process; that it will cause changes in the environment that become self-sustaining.

Carbon sinks don't all have carbon in them, but they all affect the Earth's temperature. Active carbon sinks – the earth's greenery and the seas – take carbon out of the atmosphere. Carbon stores, which include forests, the seas and also peat bogs, hold carbon out of the environment – but can be "tipped" into giving it back. And ice and snow – the cryosphere – have no carbon at all, but cool the planet by reflecting sunlight. They can be "tipped" into melting, cutting the reflective index of the entire planet. In melting, they can expose frozen peat and other stored carbon, which then melts and releases into the atmosphere. This shows how carbon sinks can work as dominoes, each contributing to the tipping of the next.

In the 2007 Consensus, runaway climate change – successive tipping of carbon sinks – is held out as a dangerous, but avoidable threat. The conclusion is that we must not let temperatures rise more than 2°C above pre-warming levels, which, the IPCC holds, could happen by about 2050. And that if we do let temperatures rise more than 2°C, carbon sinks are quite likely to start tipping. The fear is that, even if humans stopped emitting greenhouse gases at that point, successively tipping carbon sinks would cause their own warming, with humans then unable to stop further increases of many degrees.

As with sea level rises and other problems under the 2007 Consensus, this places the worst impacts several decades in the future – conveniently outside the roughly thirty-year time horizon in which major investments are evaluated, and conveniently outside the working lifetimes of those making decisions today.

But the fact is that runaway climate change is already here. The first major carbon sink to go, the North Pole's ice cover, is already doomed, and is now expected to disappear in the next two decades.

This is runaway climate change in two senses. The first is in the sense of climate change being "out of control". No rational worldwide authority would ever have chosen to allow human activity to cause enough warming such that the North Pole's ice cap would melt. Yet it is melting, and will soon disappear as a permanent feature of the Earth's landscape.

It's also runaway climate change in the sense of starting a self-sustaining process. That is, the melting of the North Pole's ice cap will cause other carbon sinks to tip, as will be described in this chapter.

So runaway climate change has arrived already. In fact, as will be described in later sections, I assert that runaway climate change began in the early 1980s.

From a wider perspective, the Earth's complement of easily accessible fossil fuels can also be considered a carbon sink – the type that holds carbon out of the environment unless disturbed. Fossil fuels were "disturbed" by humans seeking them out and burning them. Viewed as a carbon sink, the stock of fossil fuels tipped about 1910, when warming started. (Theoretically, we humans can stop our own emissions and deforestation at any time; as a practical matter, though, even slowing their rate of increase has so far proved beyond us.)

Because greenhouse gas emissions from these fossil fuels are so high, and because greenhouse gases are so slow to leave the atmosphere, the warming they cause forms a semi-permanent flame warming all the other carbon sinks. This background of warming makes it ever easier for other carbon sinks to tip, and to tip each other – as, I argue, has already begun.

In this chapter I'll demonstrate how runaway climate change has already begun by describing the major carbon sinks in some detail, then looking at the first carbon sink to tip due to warming: the North Pole's ice cap. I'll then describe the two carbon sinks this initial tipping directly endangers: northern permafrost, especially a methane-rich type of soil called yedoma, and the great northern forests. I'll then briefly describe the remaining carbon sinks, most nearly certain to tip in the coming decades, unalterably changing life on Earth.

In the next chapter, I make a pass at quantifying the sensitivity of carbon sinks to tipping – the temperature rise above which a given carbon sink is doomed – and the contribution that the carbon sink, once tipped, will make to further warming.

What are the carbon sinks?

The idea behind runaway climate change is basically a domino theory – that warming caused by humans causes some element in the environment to tip and contribute to additional warming, which then causes some other element to tip, and so on.

The analysis here will be at the level of someone examining a line of dominoes to determine whether tipping one would be likely to knock down the rest. I'll just look at the big picture here.

The actual picture is more complicated. Warming caused by humans' emissions and deforestation puts pressure on all the carbon sinks. Warming already incurred or committed to might, by itself, be enough to directly tip many or most carbon sinks, given a century or two for its effects to play out.

In addition, any change in one climate change "domino" puts pressure on all the others, as well as having specific effects on specific others. Dominoes can tip partially or completely, and at differing speeds over time.

The dominoes are different sizes, as determined by their impact on average global temperature, and have different susceptibility to tipping. However, like actual dominoes, once a climate change domino has fallen, the effects of its falling are very hard to reverse.

What I'll provide here is a first approximation, simplified in ways that tend to understate the likelihood of the dominoes actually falling, since I won't be looking in detail at cumulative or local effects. However, in doing this, I'll be describing the problem in more detail than, to my knowledge, has ever been done before – which is odd, given all the talk that has already taken place about runaway climate change as a threat.

It will take some real, difficult scientific work – which does not seem to have been begun, let alone completed, for reasons I don't understand – to assess the size and susceptibility to tipping of each carbon sink, to determine the linkages between carbon sinks, and to assign numeric probabilities within and among different tipping or non-tipping scenarios.

The major carbon sinks actually fall into three groups, two of which overlap.

The first major type of carbon sink doesn't have any carbon in it at all, but does have a strong cooling effect on the Earth. This type of carbon sink is the snow and ice that covers, on average, about 10% of the Earth – collectively called the cryosphere. (The root "cryo" means "cold" or "freezing" and is also found in words like cryogenics.)

Most snow and ice cover is in the Northern Hemisphere, so the area of Earth covered by snow and ice drops greatly during the Northern Hemisphere summer and increases greatly during its winter.

Snow and ice cover cools the Earth not primarily by being cold, but by being white and shiny, reflecting about 80% of the sunlight it receives back into space. When snow or ice cover disappears, it exposes ground or water that is much less reflective – only about 20% reflective on average.

Different types of snow and ice cover can legitimately be called carbon sinks in the sense that it will be the release of CO_2, and other mostly carbon-based greenhouse gases, that causes the warming that will damage them; and in that warming unleashed by their reduction will cause other, "true" carbon sinks – ones that actually have carbon in them – to be damaged.

The cryosphere can be divided up into different carbon sinks, which we'll discuss in more detail later in this chapter. They are:

- **Snow cover on land.** Snow cover, which alters radically with the seasons and is, by its nature, the most sensitive of the carbon sinks to warming, has hardly been studied or discussed at all in relation to climate change. Loss of snow cover means loss of reflectivity.
- **Sea-based ice, almost all in polar regions.** The main sea-based ice is the loosely termed "ice cap" of the North Pole, which is a mixed area of icebergs with some open water, and the true ice cap of the South Pole, a substantial part of which floats on open water. Loss of sea-based ice causes loss of reflectivity as seawater is exposed, but does not cause sea level rises[4].
- **Land-based glaciers.** Land-based glaciers are found across tropical, temperate and polar regions. They include Peruvian and Himalayan glaciers which serve as water sources for large populations, the ice cover of Greenland and much of the ice cover of the South Pole, which hosts a mixture of fully land-based, partly land-based and sea-based ice. Loss of land-based ice causes sea level rises as it diminishes, but only causes loss of reflectivity when the ground underneath is finally completely exposed. (Sea level rises are hugely impactful on people but have

---

[4] This is because ice expands when it freezes. The part of an iceberg that's underwater takes up the same volume as all the ice in the iceberg, including the above-water part, takes when melted. So when the iceberg melts, all the water in it takes up the same volume below sea level as the underwater part of the iceberg did before. You can prove this to yourself with a glass of water and an ice cube.

little direct effect on runaway climate change, as they simply cover unreflective land with unreflective water.)

There are also non-carbon "carbon sinks" directly created by human activity:
- **Cities and other developed areas.** Cities absorb heat by day and re-radiate it at night. Cities keep themselves warmer than their surroundings, called the "urban heat island effect" – just painting surfaces, especially rooftops, white can lessen this effect. As the total extent of cities increases, they warm the planet further.
- **Smog.** Cloud cover reflects a great deal of sunlight. Smog acts as additional cloud cover, somewhat offsetting the effects of greenhouse gases.
- **Soot.** Soot – sometimes called "black carbon" – is very absorbent of heat. When particles of soot land on ice or snow, they can melt it with surprising efficiency. As ice and snow melt away, leaving the soot particles behind, the remaining snow gets darker, melting ever faster.

The counteracting effects of smog and soot can each be greatly reduced by legislation; with world population set to rise from 6.5 billion today to an estimated peak of 9.1 billion by 2050, and farming growing ever more mechanised, cities seem destined to grow just about regardless.

Unlike the types of not-quite "carbon sinks" listed above, "true" carbon sinks have carbon in them. But there are two different, but overlapping types of these as well. One type takes carbon out of the atmosphere, but does not readily give it back up; greenery and oceans are the main such carbon sinks. The other type doesn't absorb much carbon in the here and now – as with fossil fuel deposits, it had to have done so at some point in the past – but can emit carbon into the atmosphere if disturbed.

"True" carbon sinks are:
- **Oceans.** The oceans of the world, which through chemical processes extract CO_2, in particular, from the atmosphere. This process first inters CO_2 in deeper waters, then eventually buries it at the bottom of the ocean. About a quarter of all CO_2 emitted each year is removed from the atmosphere by this effect, greatly reducing climate change.
- **Forests.** Forests – shorthand for all the greenery on Earth – which take carbon from the air (as part of their respiration and growth) and store great amounts of carbon in wood and foliage.

Forests can have their "breathing" impaired (ie by heat stress or drought), impairing their carbon uptake without releasing stored carbon. And they can be killed in ways which don't immediately release their stored carbon (ie by logging for construction). Forest soils hold a great deal of carbon as well. This dual nature means that forests need to be treated as two types of carbon sinks – the first type reflecting their capability to absorb carbon and the other type their related, but separate potential to release it.

Greenery absorbs CO_2 at a similar level as the oceans, about a quarter of all CO_2 emitted each year. When forests are burnt, damaged or killed, their ability to absorb carbon is lost, and carbon stored in them is released, mostly as CO_2 if burnt and mostly as methane if allowed to rot. Conversely, new forest absorbs a great deal of CO_2 as it grows and a smaller amount in ongoing respiration.

Finally, there are carbon sinks that don't actively take carbon out of the air or sea, but that hold it in storage – and can release it if perturbed:

- **Permafrost**. Permafrost is permanently frozen ground. Permafrost retains snow and ice cover for most or all of the year and includes vast acreages of frozen peat, which is partly decayed foliage. When permafrost melts, it loses its tendency to retain snow and ice cover, reducing the Earth's total reflectivity, and allows frozen peat and other frozen soils to decay, which releases great amounts of CO_2 and methane.
- **Peat bogs**. Peat bogs release some greenhouse gases, mostly methane, but this is slowed by their being covered with water. If the peat is allowed to dry, is burnt as fuel or is accidentally set afire, it releases vast quantities of CO_2 and/or methane.
- **Clathrates**. Also called "methane clathrates", "methane hydrates" or "methane ice", clathrates are permanently frozen or pressurised methane deposits buried in the permafrost of sea beds or deep underwater. The rapid release of methane from clathrates would cause sharp increases in local temperatures as well as in global warming, possibly triggering abrupt climate change – rises of more than a degree C per decade, perhaps several degrees C in a year – and/or runaway climate change.

So that's eleven carbon sinks identified without even trying hard. There are three related to snow and ice: snow cover, glaciers and floating ice. There are three directly generated by people: cities and soot, both of which increase warming, and smog, which decreases it.

And there are five "true" carbon sinks that absorb and/or contain carbon: the ocean, forests, permafrost, peat bogs and clathrates.

To make it an even dozen, and as mentioned above, the Earth's fossil fuels can be considered a carbon sink – one which no natural process would release in any volume, but which the specialised human activities of drilling and mining, followed by refining, transport and burning, release all too effectively. The earlier chapters on emissions relate entirely to the unleashing, by humans, of this specialised carbon sink. (The three carbon sinks generated by people listed above – cities, soot and smog – can then be grouped alongside fossil fuel emissions as being under the direct control of people.)

Even this quick survey should hint at some of the complexity of understanding carbon sinks and their interactions – and how easy it might be to create a situation in which they tip into less reflection of sunlight, less carbon absorption and/or more release of carbon into the atmosphere, each affecting the other in a runaway fashion. My argument is that one of them – the North Pole's ice cover – has, in fact, already tipped; and that all or nearly all of the others are certain to tip, in one order or another, over the coming decades, meaning that we are indeed already in runaway climate change.

The North Pole's ice cover

The North Pole's ice cover is the first carbon sink to be tipped by global warming, and the first domino to fall in runaway climate change.

Compared to 1970, the last point at which it's believed to have been relatively unaffected, the North Pole's ice cover is today less than half its former thickness and about half its former summertime extent. It experienced record shrinking in the unusually warm summer of 2007, then nearly matched that in the cooler summer of 2008. Also in 2008, sea passages all around the Pole opened up simultaneously for the first time ever recorded.

Scientists, who only recently were expecting the end of permanent sea ice – that is, nearly ice-free waters in summer – at the Pole to occur late in this century, are competing to bring their estimates forward. A straightforward projection of current trends seems to put the date at around 2030, but such projections keep proving too conservative.

Most people have not really absorbed what a huge change this is. A major geophysical characteristic of the Earth is doomed. Several important species of wildlife and the way of life of native peoples are threatened or doomed along with it.

Figure 6-1 shows the change in the extent of sea ice at its annual minimum, usually reached in September. The lines shown are the average minimum for 1979-2000 (a customary but odd choice, as melting had already begun before 1979); the minimum for 2005; and the minimum for 2007.

Figure 6-1. North Pole sea ice extent for recent years

The only ones who have really figured the implications of this out are the Russians, whose rights claims are described in the previous chapter; shipping companies plotting to save millions from quicker transits; and tour ship operators promising breakthrough polar cruises in upcoming years.

One of the key human tendencies that climate change exposes over and over is what we might call the Methuselah fallacy: the tendency not to notice changes that take place over decades or centuries. This is the human version of the "boiling frog" problem mentioned in Chapter 3.

The accelerating collapse of the North Pole's sea ice is an extreme example of this tendency. Part of the reason for it might be the effect of the 2007 Consensus, as described in Chapter 3. With climate change getting breakthrough attention in 2007 – with books, movies and speeches that brought a whole lot of new information but didn't include, nor even predict, the imminent changes at the North Pole – the new bad news shortly afterward, and its significance in relation to runaway climate change, was more or less missed.

Yet it's still amazing just how little impact this event is having on people's consciousness. If you had asked scientists in, say, 2000 what would happen if emissions weren't reduced in a few decades, they might have said, "Climate change might go runaway. At some point, the North Pole's sea ice might all melt in summer. That could cause the permafrost to melt. That would release methane that would...".

Now that the North Pole's sea ice is certain to all melt in summer sometime in the next few decades, the same scientists say that runaway climate change is still uncertain, off some distance in the future. But, with the ice committed to disappearing, runaway climate change is already here.

As I'll describe below, the melting of the North Pole's ice cap is likely to produce, at a first approximation, about 1°C in global warming. This is on top of the 0.8°C already incurred and the 0.6°C certain to follow based on emissions already released – a grand total of 2.4°C. Also, we are experiencing 0.2°C per decade of warming as a result of current emissions.

The 2007 Consensus claims we can keep warming under a total of 2°C with sharp cuts in emissions by mid-century, but that assumes a sharp decrease in global emissions by 2050 – an improbability – and no effect from the loss of North Pole sea ice, nor from damage to other carbon sinks – an impossibility.

So it seems certain we'll break through the 2°C barrier, no matter what we do from here. We need to figure out how to slow this, and how to reverse runaway climate change now that it's begun.

The direct effect

Let's take a closer look at the possible direct effect on the Earth's temperature of the melting of the North Pole's ice cover.

Our main concern is the very high reflectivity of ice and snow vs. the very low reflectivity of water. As the North Pole's sea ice decreases, exposing water, reflectivity drops, so more sunlight is absorbed and converted to heat. Also remember that the sun disappears completely for up to several months in the Arctic winter, and shines 24 hours a day for a similar period in the summer. (They don't call it "land of the midnight sun" for nothing.)

In an average summer of the last 20 years, the ice cap drops from 5.9m square miles, about 3% of the Earth's 200m square mile surface area, to 2.5m square miles – just over 1%. This means that 3.4m square miles ice up over the winter, and the average ice cover while the sun is shining is the midpoint: 4.2m square miles of the Arctic sea covered by floating ice.

When we reach iceless summers, the area that ices up in the winter is likely to be, at most, the same: 3.4m square miles. This will decline to zero at summer's peak – if not before – making for an average of 1.7m square miles covered by floating ice over the summer. The floating ice is not solid. We can assume roughly two-thirds coverage by ice, the rest water.

The average area of Earth covered by black water – 80% absorbent Of sunlight – instead of by ice – 20% absorbent – increases by 1.7m square miles. The equivalent fully absorbent area of Earth increases by about 1m square miles.

This is about .75% of the equivalent fully absorbent area of Earth. So the amount of sunlight retained as heat should rise by about this amount.

The average temperature of Earth is about 280°C above absolute zero; only about 10°C of this is from internal heat, 270°C from sunlight. So, to make a simplistic approximation, the temperature increase of Earth due to the decreased ice cover should be about 2°C.

However, on average, cloud cover is about 50%. So the average expected contribution to global warming of this change is 1°C – 1.8°F.

Much more than this amount of warming will happen locally, as the warming is occurring there first, and as global warming tends to affect the poles, and the North Pole in particular, at more than double the global average.

This kind of "first principles" calculation is not as finely tuned as a computer-modelled simulation might be, but such projections have proved reasonably accurate in the past, starting with Arrhenius' work more than a century ago. The effect, even if it varies somewhat from this projection, is certain to be too great to ignore, as is being done currently.

The North Pole is already the area of Earth most greatly affected by warming. It also contains the most sensitive carbon sinks, which are nearly certain to be tipped by the combination of existing and new warming directly due to emissions and the falling of the North Pole's ice cover domino.

The yedoma erupts

So does the melting of the North Pole's permanent sea ice constitute runaway climate change?

Looking only at the overall global temperature increase it will cause, probably. We're told that the temperature increases projected in the 2007 Consensus – which assume a previously undemonstrated discipline on humanity's part in reining in emissions – are likely to put us right at the border of runaway climate change. An additional degree or so from additional solar energy absorption at the North Pole is more than likely to be enough to tip us fully into runaway climate change.

But because of where the warming is occurring, it's even more likely. The loss of the North Pole's sea ice causes the seas to warm quickly around two other climate change dominoes – one that's becoming better-understood, the other one of the least-understood.

It was predicted by Arrhenius that polar warming due to greenhouse gas increases would be roughly double global average warming. New findings indicate that, for global warming caused by Arctic sea ice loss, the effect in the Arctic during the years of heavy loss – i.e., from now through the next one to three decades – are three and a half times the global effect, and that this effect is most accelerated in autumn. So near-future autumns may see coastal warming of 5°C in and near the Arctic Circle. This brings knock-on effects of warming – that might otherwise have been delayed for decades – forward into the here and now.

The better-understood of the two carbon sinks to be affected is permafrost, most strongly concentrated in Siberia and northern Russia, then in Alaska and Canada. Specifically, a kind of soil called yedoma, frozen peat interred in ice about 400,000 years ago, is vulnerable.

If allowed to warm, it will resume its long-interrupted rotting, releasing potentially vast amounts of methane. Permafrost over peat bogs is already believed to be melting, as an uptick in methane has been detected in the atmosphere in recent years.

Warmer seas – once ice-covered but already increasingly ice-free – and concomitant warmer air temperatures are just the thing to cause permafrost near the coast to melt and begin releasing methane. Methane emitted from warming peat will tend to linger, causing further local warming.

Areas of frozen peat will tip a tiny bit at a time in a very complex process, unpredictable in detail – it's been said that fully simulating a single peat bog would take the entire world's computing power, and then some – but inexorable overall.

As described in the next chapter, the tipping of polar and near-polar frozen peat is likely to be the next and decisive step in runaway climate change, setting it firmly in motion.

Though further study is required to be certain that the disappearance of the North Pole's sea ice alone will be enough to tip the permafrost and its frozen peat, the combination of already incurred warming, future warming already committed to, the loss of the North Pole's sea ice plus what are currently runaway emissions (see Chapter 4) are virtually certain to do it. It's a matter of urgent scientific study to identify just how fast this is likely to occur.

More speculatively, the second carbon sink of immediate concern in the North is clathrates. As mentioned at the beginning of this chapter, these are formations of frozen rotting organic matter full of methane. It's believed that in some cases ice or clathrates overlay deposits of more or less pure methane – deposits large enough that oil companies have tried to figure out how to tap them safely for energy.

A research ship recently spotted methane from clathrates boiling to the surface in the North Pole's seas. With further warming – which is now guaranteed as the North Pole's sea ice finishes tipping and frozen peat follows along – clathrates may tip as well. Which will cause local warming contributing to, and complemented by, the melting of permafrost and frozen peat bogs along the same shores.

Clathrate methane releases would constitute a disaster in their own right, the extent depending on the amount and rate of methane released. They could also trigger the release of CO_2 dissolved in deep ocean waters, bringing many millions of tons of carbon back into the atmosphere.

There is apparently at least one past instance, recorded in ice cores, of a sudden upward explosion of temperature of +5°C within one or two years. This may be the result of an explosion of clathrate emissions, or of a clathrate explosion followed by release of CO_2 from deep ocean waters. It's an awful thing to have to point out, with all our other problems, but there's pretty good evidence shaping up that governments need to prepare a civil defence-type response to a sudden, sharp, crop-killing increase in temperatures. (A "clathrate summer", the opposite of the "nuclear winter" that was widely feared when some of us were younger.)

Scientists have discovered a number of mass species extinctions in the geological record that are so far unexplained. It's believed that a spiral of methane release from clathrates and CO_2 release from the deep seas may hold part of the answer to them.

The Arctic is shaping up as the frozen tinderbox that is pushing the planet into runaway climate change. As we keep holding larger and larger matches to it in the form of ever-increasing warming from emissions, we shouldn't be surprised that it's beginning to explode.

In this unstable situation, even trees aren't helping us. Forests are mostly good for moderating the climate, extracting carbon from the air through respiration and storing it in greenery. But in snowy areas, forests break up the white reflective surface, causing more sunlight to be absorbed as heat and promoting earlier snowmelt. This causes warming not only globally but locally, contributing to the tipping of carbon sinks such as ice caps, permafrost and clathrates. So new forest growth promoted by warmer Arctic and sub-Arctic temperatures may actually contribute to warming when it occurs in snowy areas.

Greenery destruction

Less serious today than direct damage to forests through deforestation, but due to rival it in effect, is the thinning and destruction of forests and the exposure of carbon-rich soils by the side effects of climate change:
- Warming and changing weather patterns put forests under stress from heat and an ever less steady supply of water.
- Warming further increases the number and ferocity of pests that attack trees, damaging or killing them.
- Parts of damaged forests rot, releasing methane.
- Fire attacks stressed and damaged trees, destroying large areas and pumping CO_2 into the air.
- Rather than fully re-growing, forests are replaced by less carbon-dense types of forests, scrubland, grassland and so on.

The best forests for sequestering carbon are the oldest; we are destroying these rapidly and often not even allowing new, less effective forests to, literally, take root.

The scenic ski resort of Vail, Colorado is just one example. It's surrounded for miles around by trees that have been killed by beetles and are now prone to wildfires. So many trees are damaged, there's no economic use for all of them, so they're not harvested. Fire fighters can only clear firebreaks and hope for the best. When the forest eventually burns, no one knows just what will replace it.

The Amazon rainforest is another example. Destroyed in some areas, thinned in others, and warmed throughout, it's becoming more and more prone to fire. With 15 times humanity's annual carbon emissions stored in it, burning of chunks of the Amazon rainforest alone could come to rival humans as a source of greenhouse gas emissions in some years.

Forests are normally a CO_2 sink but are also capable of being a CO_2 source, releasing vast amounts of carbon as people clear land for crops or construction. Forests and other green areas don't only hold carbon directly in trees, vines and other live material, but in the rich soils underfoot. Soils can start releasing more greenhouse gases as temperatures rise, but, as with peat, this is so complex that it's hard to estimate accurately. But when land is cleared, its soils are activated as a carbon source.

With human population growing fast, and demand for livestock pasture, roads and other land-eating human needs growing even faster, people are demonstrating the ability to tip this particular climate change domino directly, through deforestation.

One example is the current assault on tropical rainforests, seen most dramatically in Indonesia. Global demand for livestock continues rising, so forests are cleared for grazing land – shades of Genghis Khan and China's cities.

There's a specific need for palm plantations – palm oil is a newly favoured replacement for trans fats, which have recently been found to be unhealthy. And palm oil is a biofuel. Palm plantations grow particularly well in area that's currently rainforest – so a vitally needed resource against climate change is destroyed in an effort to make less carbon-intensive fuel.

In Indonesia, areas covered by rainforest include extensive peat bogs. So a particularly damaging sequence occurs. With global agribusiness corporations waiting to be their customers, farmers put rainforest to the torch. This wipes out precious wildlife, directly emits CO_2 from the burning forest and, worse, releases great quantities of CO_2 and methane from soil and, especially, peat bogs embedded in the forested area.

And then – especially in unusually warm and dry periods, which of course will become more frequent with climate change – areas well beyond the initially targeted area burn. More dead wildlife, more greenhouse gases from burning forest, more greenhouse gases from soil and peat bogs.

In 1997-1998, over the Indonesian summer, purposefully set and out-of-control wildfires burned large areas. In that period, almost one-sixth of all greenhouse gas emissions worldwide came from this burning in Indonesia alone. Similar dynamics, though generally with less peat involvement, are likely across the equatorial regions of the world.

The old saying is that "two wrongs don't make a right". Here we have two rights – cutting out trans fats from people's diets and replacing fossil fuels with biofuels – creating a great and irretrievable wrong. Such unintended consequences of our efforts to reduce greenhouse gas emissions are particularly damaging and demoralising, and demand much better planning before new laws and regulations are passed.

Greenery tends to grow faster with more CO_2, so forests, like oceans, are not only taking out about a quarter of the CO_2 emitted each year; they increase their uptake, though not enough to keep up with the increasing pollution, as the "pressure" of CO_2 increases. But, like oceans, they have their limits; research shows that most plants sharply decrease their CO_2 absorption past a certain point of warming, which may be exceeded for most plants in a decade or two of continuing increases in temperatures.

The combined effect of our direct assault on forests and their decline and disappearance as a side effect of warming and climate change is likely to be a series of collapses of forest ecosystems.

Yet this cycle of decline, leading to warming, leading to further decline, leading to more warming has not been carefully studied. While the 2007 IPCC Report does project that some forested areas will degrade to savannah, there's not a chapter on fire in the entire document.

Forest destruction by the combined effects of deforestation and warming is likely to quickly erode, then virtually destroy half the environment's ability to absorb CO_2, while throwing more and more of the total 280 gigatons of carbon tied up in forests today – more than the total amount currently in the atmosphere – into the air as well.

Other carbon sinks

What is the state of other carbon sinks? How close are they to tipping?

The Greenland and South Polar ice caps are of huge concern to humans, because if they melt – and Greenland, in particular, is off to a good start – they'll raise sea levels by many metres each, hugely disrupting human life.

There are two factors that may make even these huge blocks of ice tumble quickly. The first is the kind of accelerated warming from tipping carbon sinks described in this book, which is multiplied at the poles. The second is ice melting dynamics that may cause even large masses of ice to crumble quickly once they begin to melt.

Scientists have been surprised by the way in which the largest land-based glaciers are responding to warming. Pools of water form on the ice surface and, despite the pools being only slightly above freezing themselves, cause rapid melting on their undersides, so they bore right down through the ice.

The original small pond on the surface of the ice escalates into a kind of waterfall in the ice through which increasing amounts of water run.

Once the water melts through to ground, it runs downhill, melting the ice at the interface where the glacier is anchored to the ground.

This transforms an initially "sticky" ice/ground connection into, literally, a slippery slope, down which the ice can start sliding toward the sea.

At this point the only thing stopping the ice from flowing is the ice downhill from it. Each new block of ice that falls into the sea – beyond the normal degree of "calving", as the process is called – puts all the ice behind it at greater risk. In some areas, ice that had formerly acted like a cork in a wine bottle has given way, freeing the ice behind it to flow more rapidly to the sea.

If either increased warming or these formerly unknown ice melting dynamics turn out to be significant goads to melting – and there's active scientific disagreement on this – enormously disruptive sea rises of a metre by mid-century may occur.

The South Pole has had a degree of immunity to climate change so far. Unlike the Arctic Circle, which is extensively penetrated by continental land masses running north and south, the Antarctic is centred on its own nearly circular continent and is surrounded by a belt of cold winds and cold ocean currents. This belt is evocatively called the Antarctic Wall.

Although the Antarctic has suffered some notable ice loss and warming, with widely discussed threats to penguin populations, it's also offset some effects of warming elsewhere. Snowfall in Antarctica has increased even as ice collapses occur on the fringes. Antarctic sea ice has expanded to make up some of the gap created by the steady melting of sea ice in the North.

But in a warming world, these feedbacks can only partly offset the loss of ice and snow cover. The areas of the world in which snow and ice can "stick" (to restore reflectivity) and accumulate (to keep sea levels lower) decrease steadily.

Just how quickly the Antarctic barrier is breached, and how extensively, will make a difference of perhaps decades in major questions such as the rate of sea level rise.

In a sense, though, it doesn't matter exactly how fast the ice is expected to melt. It's immoral of the current generation to accept a future of more or less rapidly melting ice and a permanently damaged civilisation – whether it's ourselves, our children or their children who directly bear a given foot or metre of the overall sea-level rise.

As far as contributing to runaway warming, sea level rises are unimportant; all that matters is how much shiny white surface area snow and ice present to the sun. Reflection continues until all the ice in a given area is gone, exposing the water or land beneath. Even the most pessimistic estimates put the loss of all surface area, for large expanses of currently thick glaciers, many decades in the future.

Finally, in looking at carbon sinks, the oceans are a carbon sink in that they absorb CO_2 from the atmosphere through a chemical process. This process takes up about 25% of annual CO_2 emissions, crucially important in slowing climate change.

The oceans act as a negative feedback loop on CO_2 emissions: the more CO_2 that goes into the atmosphere, the more oceans absorb. But it's never more than a proportion – it reduces the problem to a certain extent, but will not keep up fully with increases in emissions.

There's evidence that absorption of CO_2 by the oceans is slowing. If this progresses, it is the tipping of a crucial carbon sink.

There is also a danger, as mentioned above with reference to clathrates, that some process or another will begin to unleash the CO_2 interred in the deeper layers of the ocean. This is a truly frightening prospect, and is one of those carbon sink tippings that could rapidly change the world. However, it's not well understood; it needs research, as does the entire portfolio of carbon sinks.

Chapter 7. A Model for Runaway Climate Change

> In this chapter
> - What a model accomplishes
> - Filling in the model
> - Long-lasting effects of warming
> - The impact of losing carbon sinks

Don't let the phrase "model" in the title of this chapter scare you. I'm a wannabe scientist, rather than an actual one, so this chapter is highly likely to be comprehensible to you (or it wouldn't be to me).

It's important to humanity that scientists have a precise understanding, and the rest of us have a rough understanding, of just what runaway climate change means. That includes how we'll know if it's occurring, how it will play out as and when it occurs, and whether it might be stopped at some point short of completion once it has begun. (I'm suggesting that this last answer is likely to be "no" by using the term "runaway" as the title of this book.)

We can build a model to represent most of the aspects of runaway climate change. The model will help us understand whether or not it's happening and how it's likely to play out. And the model can be improved with further research and discussion to enhance our understanding.

In this chapter, I'll describe how I've built up such a model, what it seeks to accomplish and what its strengths and weaknesses are. The completed model is shown in Figure 7-1 below.

I don't argue that the model is definitive; the research has not yet been done to make that possible for anyone to accomplish. But it represents the current state of research reasonably well; indicates that, if current understandings are accurate, we are in runaway climate change; and provides a framework in which to complete the analysis needed for a definitive answer to the question.

How the dominoes work

In the model I characterise each of the world's carbon sinks – snow and ice cover, forests and so on – as a domino. For a domino to "tip" means that it's committed to eventual collapse, as the North Pole's ice cover is today. For a domino to collapse means that its cooling effect on the Earth – through reflectivity, by absorbing carbon or by keeping carbon stored out of the environment – is nearly or completely lost.

The domino's height is the amount of warming it adds to the environment when it's in its peak period of impact.

In the model, I arrange the dominoes along a line by how much warming is required to cause them to tip. For instance, the North Pole's ice cap – whose height indicates its contribution of approximately 1°C to warming - seems to have tipped at 0.3°C of overall warming. Two dominoes that tip at the same amount of warming will be placed at the same position on the line.

The line also has an indicator that shows the current degree of warming, from 0°C on the left end to +10°C on the right. The indicator is currently at +0.8°C and shifts to the right as warming proceeds.

The model is a snapshot in time and changes as warming proceeds. The indicator can potentially shift leftward as well as rightward. For instance, smog, as we will discuss later, has a cooling effect; if smog got more intense while sources of warming stopped, the indicator would shift leftward – as may have occurred in the 1950s.

Methane "only" lasts about a dozen years in the atmosphere – a long time if harvests are failing every year the methane is present, or if it helps tip another domino while active – so a methane release can shift the temperature indicator rightward for more than a decade as it causes additional heating, then let it slip gradually back to the left as the methane breaks down and the heat slowly dissipates.

To roughly indicate the degree to which a given carbon sink's effects last or fade with time, I use fully black dominoes for permanent warming effects, darker grey for CO_2 and light grey for methane.

With these rules in place, "all" that remains – a difficult task subject to much uncertainty – is to characterise, or describe, each carbon sink as a domino. We need to determine its sensitivity – the temperature at which it tips – and its impact, the peak contribution it will make to warming. These determine the domino's placement on the line (sensitivity) and its height (impact).

To determine whether climate change is truly runaway, under a strict definition of the term, is done by excluding any and all future human greenhouse gas emissions from the model, then assessing whether dynamics already in motion will tip existing carbon sinks.

It's possible – though highly speculative – that a given geo-engineering effort could offset or even counteract the impact of the collapse of a given carbon sink. Such efforts can be assessed by pencilling them in as dominoes that extend below the line, somewhat offsetting the positive dominoes to some extent.

(The model could also be used with a domino to represent anticipated human greenhouse gas emissions. However, any reasonable current projection of these is so large that this domino would tip over most of the others – and, as we shall see, the other dominoes are large enough, in sum, to tip one another over without the need for any further human greenhouse gas emissions. Clearly, radical efforts are needed in both emissions reductions and in using geo-engineering efforts to save threatened carbon sinks.)

In a sign that the 2007 Consensus is breaking down, some scientists are saying that the net human contribution must go negative – that is, emissions must drop sharply and be outweighed by geo-engineering – to prevent runaway climate change. This is expressed as reaching an atmospheric CO_2 level somewhat below today's, but really represents a variety of emissions reduction and geo-engineering efforts that need to be assessed separately – and represented as separate, negative dominoes.

An initial version of the model, with estimated values, and excluding future human greenhouse gas emissions and deforestation, is shown below, after a brief introduction. There's much disagreement as to the size of various carbon sinks, partly due to a lack of study and partly due to a lack of discussion among contributors with differing perspectives. What the model makes clear is that, as far as we can tell from today, we are already in runaway climate change; the dominoes have no problem tipping each other over. So the tipping of carbon sinks appears to already be self-sustaining.

One thing this model leaves out, which makes things worse, is local knock-on effects as opposed to global ones. For instance, the disappearance of temperate zone glaciers worldwide, spread over one to two centuries, has a significant and permanent effect by cutting reflectivity. But it's broad and gradual enough not to have a disproportionate impact on any other specific carbon sink. The whole world just gets slowly warmer.

The erosion of the North Pole's ice cover, on the other hand, has disproportionate local effects on frozen areas in the North, including large areas of carbon-rich yedoma (frozen peat) and Northern forests. Methane releases similarly may have a local effect. Modelling these local effects approximately will require a great deal of work; modelling them in detail is beyond our current capabilities.

And if you're sharp, you may recognise a simplification in the model: a given carbon sink doesn't actually just fall over like a domino at a given point of warming. Instead, the rate of emissions or other change responds to the local temperature, which interacts with global temperature, both of which are changing over time. It would be more accurate to render a carbon sink's effects as a more or less bell-shaped curve rising and falling over different degrees of warming than as a single domino.

In a warming world, though, the dominoes are a good first approximation – and since our data is also approximate, given that carbon sinks have not been carefully studied nor modelled, simple dominoes may be more appropriate at this point than complex curves. I look forward to the day when someone smarter than me gathers the necessary data and renders the dominoes chart as a set of stacked curves instead.

What a model does for us

Now that I've described how the model is constructed, but before actually putting together an instance of the model with some initial guesstimates, it's worth considering what we hope a model will do for us.

The next thing the model will do is tell us just how runaway climate change is – how much room, if any, remains for further emissions from humanity. If runaway climate change really is runaway, the dominoes, once the first is tipped, will tip one another – without further intervention. Our model will help us know where we stand.

The first thing we'll look for is any gaps between dominoes – where the tipping of previous dominoes doesn't take us to the next one. This will tell us if there are any "firebreaks" – plateaus where it would be easier to stop runaway climate change once it had begun.

If there are no firebreaks, we know that runaway climate change is just that – unstoppable by any previously known means. So previously unknown means, under the overall label of "geo-engineering, must be considered.

Perhaps the most important effect of the model is to begin to convert the ideological debate on climate change into a series of fact-finding exercises. Few climate change sceptics would claim that losing the Amazon rain forest would neither indicate – nor cause – problems. Few activists would claim they know exactly how endangered the Amazon rain forest is. What the model does is convert vague and ultimately unanswerable questions about who's right and who's wrong into scientifically answerable questions about what, how much and when.

That's "when" as in "at what point in warming". Another way this particular model will help lower the temperature in ideological debates is by staying clear, in an initial discussion, of specific dates, a potentially emotional topic which can generate more heat than light if taken on at the beginning. The dates discussion can be deferred until one spells out the assumptions and estimates going into this model; one can then "play" an instance of the model out over time to see what its effects might be at any specific date.

In addition to timing, this model also leaves out effects on sea levels, which have the most immediate and direct effect on humanity, but don't much affect whether climate change is runaway or not. Sea level rises can be estimated as an output once an iteration of the model has been specified.

Filling in the model

To fill in the model, I need to estimate the sensitivity of various carbon sinks and their total impact on temperatures. I also need to specify whether the loss of the sink causes lost snow and ice cover, which has a permanent effect; CO_2 releases, with a half-life of about 100 years, which I'll call long-term; or releases of methane, which breaks down after an average period of about 12 years, which I'll call short-term.

To begin with, I need to account for warming that's due from the emissions we've already placed in the atmosphere and the deforestation that's already taken place. It takes time for the full effect of emissions to be felt, so – in addition to the 0.8°C of warming we've already experienced – we're due an additional 0.6°C of warming just from the direct effects of past emissions and deforestation, even if new emissions and deforestation stopped today. Because it's mostly due to CO_2, this is considered long-term.

Now on to carbon sinks, the main area of concern for runaway climate change. The first major carbon sink to tip has been, I assert, the North Pole's ice cover. (Given that humanity ignored the clear warning given to us by the decline of most tropical- and temperate-zone glaciers.)

Based on the rate of thinning of the North Pole's ice cover – half gone in 40 years – it tipped early in the 1980s, I calculate. (The tipping point is estimated as one-sixth of the way to complete loss.) This puts the tipping point for the North Pole's ice cover at the 0.3°C mark of warming, and it will have an estimated impact of about 1°C. I'll colour it black to indicate that it's a permanent change, a loss of reflectivity we won't recover until the climate somehow changes again. (And yes, I recommend that we contrive to cause that change so the ice comes back.)

The next carbon sink to place is the emission of methane from permafrost. The impact of this at maximum depends on the amount of methane buried and the rate at which it sublimates from frozen to airborne, before being broken down by sunlight an average of a dozen years later.

A UCLA hydrologist has estimated vulnerable Siberian methane deposits at 70bn tons of carbon. Smaller Canadian and Alaskan deposits are also at serious risk. If the Siberian deposits alone were to deplete over the next 100 years, they would add an average of 1Gt of CO_2 equivalent into the atmosphere in peak years, roughly matching current emissions from wetlands and agriculture. (There are other, higher estimates which need to be considered as well, but I'll use this relatively low one here.) This would add an estimated 10% to 25% to global warming.

And emissions from the permafrost cause a doubly positive feedback: they increase global temperatures, causing permafrost emissions to increase further, and they increase local temperatures even more, causing permafrost emissions to increase even more than that. So the amount of carbon emissions from the permafrost would likely increase with temperature, causing their share to stay roughly the same as emissions increased until largely depleted.

It's very hard to predict just what emissions from the permafrost will be across a range of scenarios. In a rapidly warming world, permafrost emissions might contribute about 0.5°C to temperatures at or near their peak, which will extend over several decades. So I've put the height of the domino at 0.5°C and coloured it light grey to indicate that it's made up of methane, with its relatively short lifespan.

But where to place it? Increasing methane levels in the atmosphere indicate that the permafrost has begun to melt in the past few years. The collapse of the North Pole's ice cover can only increase its speed of melting – and this doesn't require further emissions nor temperature increases, just a few decades of time. This still leaves a small margin of uncertainty, so I suggest placing the permafrost domino at the 1°C mark, just past today's 0.8°C, pending further research.

This still leaves three-quarters of the world's buried methane unaccounted for. It is susceptible to being released by permafrost warming, desertification and other causes. Because this has not been carefully studied, we can only estimate how much will be released and at what point of warming.

It seems safe to say that a third of this remaining methane will be released as the Arctic and other regions undergo further warming. So I've placed another bar with height of 0.5°C and coloured light grey at the 2°C mark. Further research will be needed to place two more bars of similar height to indicate the release of the remaining 50% of the world's buried methane.

Clathrates, likely to be affected by the permafrost domino, are a real wild card, especially given their potentially catastrophic impact of perhaps 2°C or more of warming over many years. I've left them out of the model, but they urgently need further study. Clathrates, possibly accompanied by carbon release from the deep seas, could be like a bomb going off in our model, knocking over dominoes all down the line.

I'll do the rest of this analysis in broad brush strokes because of the uncertainties involved. The Earth's remaining forests absorb about a quarter of our emissions today. If we were to stop cutting them down tomorrow – which can only be a supposition, as it's unlikely to happen – then the remaining danger to them would be from warming.

There's a point at which warming will choke, kill or burn enough forest to offset the absorption being done by the healthy portions, which will move them from a net absorber of a quarter of our emissions to zero effect.

We consider this impact permanent as an accounting practice, because then we can consider the positive impact of planting new forests as permanent as well. This separates something we can't control, the destruction of forests' carbon-fixing capability as a side effect of warming, from something we very much can, the planting of new forests.

A report presented at the United Nations Forum on Forests (UNFF) in spring 2009 says that forests are likely to stop absorbing carbon entirely at total warming of 2.5°C. So I put forests tipping to carbon neutrality at 2.5°C; since they today absorb one fourth of our carbon, the impact at that point of forests moving from helpful to neutral would be 0.5°C. This domino can be coloured black because it's permanent, to be offset by a domino pointing downward, to indicate cooling, if we reverse temperature trends and forests resume absorbing carbon.

There's also the effect of the forests becoming a carbon source, not just the absence of a carbon sink. This is if they spiral toward overall destruction. The UK's Meteorological Office recently projected that 40% of the Amazon rainforest would go even if emissions stopped by 2050; such a halt is very unlikely to happen, and other carbon sinks are tipping, so this estimate can be taken as an indication that the world's forests as a whole are on a hiding to disaster.

The Amazon rainforest alone is estimated to contain carbon equivalent to 15 years' worth of current emissions. If we assume that half the world's remaining greenery could go at, say, 3.5°C; and that it contains about 30 times today's annual emissions, and will take 30 years to burn or otherwise die off; then we have an impact of about 1°C. This impact is dark grey because it's mostly CO_2 from burning, while rotting (rather than burning) forest remains emit methane.

It's not possible to usefully estimate what would happen to the remaining half of forests due to warming without detailed scientific study. A fair amount of forest could gradually shift northward and, to a lesser extent (as there's less land available), southward, but how quickly losses could be partly offset by such gains is unknown.

The effects of this are not all beneficial, as adding trees to the remaining snow- and ice-covered areas disrupts their reflectivity – carbon gets stored, but sunlight that was formerly reflected gets absorbed as heat as well. So I've left the remaining half of today's forests out of the model.

Also note that this is only warming-caused damage to forests; we are accelerating this process by clearing land directly, which means we could see the same warming from forest destruction happening sooner, more completely and in a more compressed way that has a greater peak impact on temperatures.

Also note that we can divide the total impact of greenery by region, for instance giving the Amazon rainforest, the forests of the American Southwest and the Siberian forests their own, smaller bars, each the appropriate height, to gain greater resolution in our estimates.

This leaves the oceans. Like the forests, they take about a quarter of our current emissions, an uptake which is showing signs of dropping as warming continues. At a high enough point of warming, intensified storms may "turn" the upper layers of the seas enough to expose interred carbon more or less steadily over time, offsetting the seas' remaining carbon-fixing capability and making them net neutral.

We can divide the seas' net carbon-fixing capacity into halves for easier processing, and assume we lose half of it at 2°C and half at 4°C, both permanent. (To be offset by downward-pointing bars if we can bring temperatures down again and see the ocean's capacity for absorption resume.) The part occurring at 2°C would be "worth" an eighth of the impact of then-current carbon emissions, about 0.25°C; the part occurring at 4°C would have an impact of about 0.5°C. Again, these are estimates, and need both further study and further subdivision – say, by more accurately estimating the impact of each degree of warming from other sources on the seas' capability, and giving each degree's impact its own domino.

We can look at some further losses that are more remote in time because they play out gradually, but perhaps not that far away in terms of additional temperature required to doom the dominoes involved. We may lose a third of the remaining ice- and snow-covered surface area (since snow won't "stick" on hotter ground) at 2°C of global warming, remembering that warming is twice as great in polar regions as the global average that 2°C represents. This loss of reflectivity would cause overall warming of perhaps 1°C, similar to the loss of the North Pole's ice cap, coloured black as it's permanent.

The remaining snow-covered surface area doesn't disappear until the great ice caps of Greenland and the South Pole disappear completely. The ice caps, and their surface reflectivity, may not completely disappear until 3°C (Greenland), 4°C (the floating part of the South Pole) and 5°C (the land-based part), respectively – perhaps a century after each of these temperature landmarks is reached, but our model doesn't account directly for time. Greenland can be considered to account for about 1°C of total warming and the South Pole's two parts for about 1°C each, and each such domino is coloured black.

If, after further study, one domino barely reaches the next, there might be an opportunity to create a firebreak – for instance, by creating some kind of artificial carbon sink while, for instance, a methane eruption is given time to have its effect, then dissipate.

We can also look for places where dominoes are all too near each other and study the time element more closely, looking for spikes in temperature due to increases in warming from two or more sources at once. Such periods may not be easy years in which to be alive.

The model gives us a way to assess the need for, and potential effect of, various possible interventions. For instance, the reflective effect of snow could be replaced by putting carefully chosen particulates – a sort of health-neutral smog – in the skies. (Sulfur particles have recently been recommended for this purpose – talk about hell on Earth!) We can place these interventions in the model as offsets against the tipping of carbon sinks, extending below the line to indicate their estimated cooling impact.

The model could also be updated with varying estimates of human-caused emissions at different levels of warming, but this has a time dependency that complicates the model greatly, and would not help assess runaway climate change against the strict definition of it I'm using here. It could also be updated with various interventions to offset warming, such as a massive programme of reforestation, which again would have a time factor but less so, as the greenery would be permanent – if it was effective in helping stop further warming and thereby protected itself from degradation and destruction.

So to sum up our guesstimates, we have the following changes from the loss of ice cover, forests and the ocean's ability to absorb carbon:
- Warming already experienced due to humans' greenhouse gas emissions is 0.8°C of impact – long-term
- Warming still due from current emissions, 0.6°C of impact, beginning at 0.8°C – long-term
- North Pole ice loss, 1°C impact, tipping at 0.3°C – permanent
- Permafrost loss, 0.5°C impact, tipping at 1°C – short-term
- Loss of one-third of remaining buried methane, 0.5°C, tipping at 2°C – short-term
- Forests going neutral, 0.5°C, tipping at 2.5°C – permanent
- Loss of half of remaining forest, 1°C, at 3.5°C – long-term
- Ocean uptake loss, .25°C heat increase at 2°C of warming and 0.5°C at 4°C – permanent
- Snow-covered surface area loss of 1°C each at 3°C, 4°C and 5°C – permanent.

The complete model is shown in Figure 7-1. Total warming above pre-warming norms, between 0°C and 5°C of total temperature rise, is 8.65°C, with no gaps between dominoes.

An additional 1°C of relatively short-term warming from remaining buried methane remains to be placed, as does 1°C from the loss of the remaining half of today's forest extent, and the potentially very large impact of clathrates. There are many additional uncertainties in the model as shown. Water vapour is a greenhouse gas and gets more potent with temperature – but it may contribute to additional cloud formation (it doesn't seem to have done this so far), which is a negative feedback, so this needs further study.

There may be additional positive and negative feedbacks; it seems that there are more possible positive feedbacks than negative ones below 5°C of additional warming, whereas above 5°C, there could be negative feedbacks that slow or reduce the total heating, but probably by only a fraction of the total. Yet there are also potentially explosive unknowns among positive feedbacks at higher degrees of heating as well.

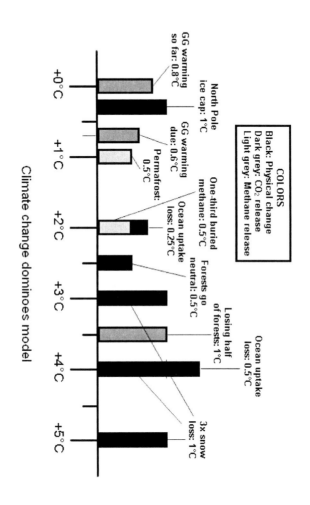

Figure 7-1. Climate change dominoes model

Remember that climate change is only "runaway", in the strictest sense, if the dominoes tip each other without any gaps, and without any further emissions or deforestation from people. So let's play this instance of the model out over the range of temperatures:
- Warming that's already occurred (0.8°C) plus warming that's still to come from existing emissions (0.6°C) takes us to 1.4°C total above pre-warming conditions;
- Warming of 1°C from loss of the North Pole's ice cap, already committed, takes us to 2.4°C total;
- At 2.4°C we are well past the warming required for loss from the permafrost (0.5°C) and one-third of the remaining buried methane (0.5°C), plus the first part of ocean uptake loss (0.25°C), taking the world to 3.65°C of total warming;
- This is more than enough to cause the first tranche of snow-covered surface area to be lost (1°C), remaining forests to go neutral (0.5°C), then a cumulative loss of half the world's forests (1°C), bringing total warming to 6.15°C;
- Well before 6.15°C, the remaining effects will kick in: the loss of the ocean's remaining carbon uptake capability (0.5°C) and the loss of the remaining two tranches of snow- and ice-covered surface area (1°C each).

The total is 8.65°C – runaway climate change, with no useful gaps, even with much methane buried in soils, substantial forest loss, clathrates, carbon release from the deep seas and water vapour impact excluded.

A quick look at timing shows that most of the impacts are likely to take place in this century: current and already committed warming from past emissions (1.4°C total), warming from loss of the North Pole's ice cap (1°C), the first half or more of a century of ongoing methane emissions from permafrost loss (0.5°C), the loss of one-third of remaining buried methane (0.5°C), forests going neutral (0.5°C), the first part of ocean uptake loss (0.25°C), loss of half the world's remaining forests (1°C) and loss of the first third of snow-covered surface area (1°C). This is the first four bullets above, and they total 6.35°C of warming.

This approximately 6°C of warming is in addition to warming caused by human greenhouse gas emissions and deforestation, which in the most optimistic plans may be held to 2°C, but is currently on track to exceed 7°C in this century (see Chapter 4). I briefly address what this may mean at the end of this chapter.

The permanence of effects

Of the effects described, only the permafrost loss is short-term, being made up largely of methane. Although it's only short-term at the end: the methane releases, though any given release lasts only twelve years in the atmosphere, will be spread out over something approaching a century, keeping the effect in play for a long time. (Methane releases from clathrates might be short-term in the same sense – the methane breaks up in little over a decade, but the releases continue over an extended period – as well; or they might be truly short-term, a quick explosion over a very few years.)

If the dominoes were far enough apart, and other warming could somehow be stopped or even partly reversed while the no-longer-"perma"frost was defrosting, permafrost loss might not tip other dominoes. But unfortunately, even if the permafrost domino were removed entirely, the other dominoes would still be sufficient to tip one another over.

The "bloom" of methane from melting permafrost has two effects. In terms of runaway climate change, it brings forward the destruction of some permanent carbon sinks such as forests and non-glacial snow-covered areas, causing fairly immediate warming from those sources. But in immediate effects, it's even worse. By adding about half a degree of warming rapidly, which – along with other, slower warming – would shift climate zones and worsen the weather as well as directly affecting crop growth, it could impact harvests in many areas quickly and with little warning. We would not have the several years needed to try to offset failed harvests in sensitive areas with new planting in the currently undeveloped areas that would be warming sufficiently to become viable.

Until recently, we might have hoped that the CO_2-based dominoes might not be too long-lasting, and perhaps subject to some kind of manipulation. But recent study has found that carbon already hangs around much longer in the environment than previously understood. Also, carbon uptake by nature is already slowing – and, as the model shows, due to nearly stop. This means that the CO_2-based dominoes are unlikely to be able to be isolated from other dominoes.

The warming caused not only by CO_2, but by all greenhouse gases is also persistent. It "stays in the system" for many, many years – perhaps centuries. The heat has nowhere to go except to gradually radiate back out into space, and that's largely blocked by accumulated emissions. Our descendants will be living with this heat for perhaps a millennium after we're gone.

Of course, humanity is hardly sitting and waiting to see if runaway climate change happens, much less fighting it – we're rushing it along. Humanity's own emissions are accelerating upward, on track to guarantee every bad result described above. And on current plans, the forest neutralisation and destruction that's seemingly somewhat remote in the model, at 2°C and 3°C of warming respectively, is happening at today's lower temperatures due to our direct clearing of land.

Regardless, thanks to the model, we now have an initial answer to our question as to whether there truly is runaway climate change: yes. Further study will certainly shift specific impacts and sensitivities. But overall, within our current rough understanding of what's out there in the environment, climate change is runaway today.

Due to the effects of losing the North Pole's ice cover, already underway, there is no firebreak up to the 2°C mark, as stated by the IPCC and others. If we stopped destroying forests today and stopped emissions today – when we're already at 0.8°C of warming – we would still incur 0.6°C of additional warming already committed due to past emissions and 1°C of warming due to the melting of the North Pole's ice cover. That takes us to 2.4°C – past the supposed firebreak – at which point we suffer 0.5°C of warming from forests going neutral plus 0.25°C from oceans going neutral, taking us to 3.15°C, and on and on.

Still, the model is helping us here in how we think about these challenges. We can separate permanent effects from long-term ones from short-term ones, and examine the "gearing" between them clinically – temporarily setting aside hugely important issues of timing and survivability – to get a clear overall picture.

Once we agree on a working specification of the model, including understanding where uncertainties remain, we can then identify and prepare for immediate challenges – most importantly, protecting each year's harvest; medium-term challenges, measured in decades and small numbers of centuries; and ongoing challenges, measured in centuries and millennia.

We can also add human-generated emissions into the model, accounting for different scenarios for changes in emissions over time. This will be the beginning of planning how to survive the impacts of climate change with as little loss of life as possible.

What is the total impact?

If you add together the likely path of human-caused emissions and the tipping of carbon sinks described above, what's the total impact likely to be?

We've seen in Chapter 4 that human-generated emissions are likely to cause roughly 7°C of additional warming during this century. And that warming already incurred is almost certain to tip carbon sinks, adding an additional 6°C in this century alone. So the total in this century is about 13°C, followed by about 3°C in future centuries as the world's remaining ice caps melt.

One leg of this – the human-generated 7°C – is subject to involuntary as well as voluntary reductions. The most aggressive plans for voluntary reductions seek to keep the direct human contribution to warming in this century to 2°C. Even if these voluntary reductions fail, enough environmental disruption may occur so as to kill enough people, and reduce the technological state of those who remain, so as to reduce the total "contribution" by humans to warming below the expected 7°C.

What do others have to say about this? The well-respected Hadley Centre, part of the UK Government's Meteorological Office, predicts a "worst-case" rise of "more than 5°C" by the end of the century from emissions alone if they are not strongly controlled by 2050. However, these predictions are based on IPCC scenarios, which we've already shown are both baseless and far below the current path of emissions. Also, they do not include runaway climate change – significant feedbacks from nature, which we've shown can be expected to reach 6°C in this century alone.

Negative feedbacks that somewhat reduce the total temperature rise, from the human activities side, from nature or a combination, are certainly possible in the higher reaches of temperature increase. Increasing cloud cover is the most obvious candidate. However, this may only be a negative feedback above a certain point, say for example 8°C of warming; if the temperature started to dip below that level again, the clouds might lessen in response, helping to keep the temperature relatively high.

Of course, positive feedbacks that make things even worse are also quite possible, some of which are noted above. Scientists have not researched this – largely, it seems, because they have had difficulty believing that humanity would let things reach the perilous point we are now at. (Which we've reached partly because of complacency and poor communications by scientists.)

So, including the most optimistic scenarios for human emissions and deforestation, as well as the business as usual predictions, the conservative thing to say is that we will probably see a rise of 8-13°C in this century, with 10°C as the mid-range single number. Further study will be needed to give a more accurate estimate of any interactions involved. Of course, anything approaching 10°C of warming, especially happening so quickly – let alone anything exceeding that amount, or happening even more rapidly – will be the greatest disaster humanity has ever encountered.

Another possibility is that humanity looks at the path it's on and changes course quickly, so that real breakthroughs are achieved in eliminating emissions, stopping deforestation and fighting runaway climate change – likely to include geo-engineering – as described in Chapter 9.

Clearly, an immense effort is needed to prevent utter disaster. However, that effort – difficult enough to muster in the best of circumstances – will be even harder to put together as we head from today's economic problems to a related set of challenges, described in the next chapter, which are likely to fully absorb humanity's attention between now and mid-century.

Chapter 8. How the World Will Worsen to 2050

> In this chapter
> - The population boom levels off
> - Peak Food
> - The impact of air pollution
> - Damage to the environment
> - The direct effects of climate change
> - The side effects of humanity's response
> - Toward Hot Earth?

The worst kind of crisis is the one you don't know you're in yet. There are encouraging signs that the world is waking up to the need to cut emissions, with America preparing to take on a leadership role again thanks to the Obama administration.

But the world has yet to face the facts – that emissions will continue rising even if the developed world finally makes cuts; that carbon sinks have started tipping in nature, taking climate change out of mankind's control.

As they finally hit, the shock of these realisations will be huge, and the effort needed to even begin to confront them will be as wrenching as anything mankind has faced. But there are other trends, most of which will be worsened by climate change, that even by themselves would pose a huge challenge. These problems will cause severe impacts on humanity just as it needs to be devoting all its energies to fighting runaway climate change.

These interrelated problems, a continuation and culmination of the megatrends described in Chapter 1, are:
- Population growth, with global population expected to peak at 9.1 billion in about 2050;
- Water shortages, with water tables being drawn down to the point of uselessness for agriculture and crucial glaciers disappearing due to global warming;
- Food shortages and higher food prices, with farmland, rangeland and fisheries overtaxed even by today's needs, let alone tomorrow's;

- Pollution of the old-fashioned kind – smog, effluent runoff into the sea and so on – worsening steadily;
- A demographic crunch, with developed countries, as well as China, facing aging populations and a vast shortfall in money to support them;
- The current, possibly long-lasting, economic crisis that damages the prospects of people worldwide;
- Extinctions in the natural world;
- Early effects of climate change;
- Side effects of the response to climate change;
- The emergence of Hot Earth – a world with 10°C higher temperatures, extensive desertification and a severely damaged environment – as a realistic possibility.

Just for one example, the demographic crunch would be a big problem on its own. If population growth really does level off by 2050, it's estimated that the group over 60 at that time will represent an historically unprecedented – and unlikely to ever be repeated – one third of the world's population.

This group will be particularly vulnerable to problems caused by climate change, such as heat waves and flooding. And the cost of pensions and health care for the elderly will be the highest ever, just when societies are struggling on so many other fronts.

In this chapter, however, I'll focus on the most tightly interrelated problems directly affecting and affected by runaway climate change – population growth, water shortages, food shortages and pollution. I'll also deal briefly with the direct effects of climate change and the effects of the human response to it. Finally, I'll outline Hot Earth, the new – and nearly barren – version of our current world that we may, in effect, be forced to migrate to.

From all this, it's clear that the focus and resources to even fully take on board, let alone deal effectively with, runaway climate change will be in short supply over the coming decades.

The end of the population boom?

The world has been through panics about population growth before, and understandably so; the increase in population from less than 2 billion before World War II to 6.5 billion today was unprecedented and frightening in its implications. It is in many ways impressive that there are "only" about 850 million people suffering regularly from hunger today, and that widespread famine and large-scale wars have been avoided since the end of World War II.

Much of the solution was the Green Revolution that began in the 1940s, an application of science, technology and planning which saw agricultural productivity per acre more than double worldwide over several decades. This is a success that seems to bode well, not only for feeding even more people but as a template for the fight against climate change.

It seems logical that handling an increase projected to be "only" 2.6 billion more people to 2050, with growth then set to turn to a slow decline as people who are becoming better-off gradually have fewer children, should therefore be manageable.

The problem is that feeding even the current population is being done in a way that's unsustainable. Even without climate change, continuing to feed the current population would be a challenge, let alone feeding billions more.

And the additional 2.6 billion people expected will almost all be born in the poor and crowded countries in Asia and Africa that are least equipped to feed them and expected to be hit hardest by climate change.

As part of those concerns, though, we also need to reconsider the projection that populations will level off by 2050. This is assumed to be on the cards as most people in the world become urbanised and relatively well-off, since birth rates tend to decline as adults become better-educated and better-off.

But peoples who fall short of this level become stuck in what's called the demographic trap.

In the demographic trap, the people in a given country or region get access to basic education and health care; enough to sharply improve survival rates, but not enough to allow parents to manage their family sizes. Average family size and population shoots up. But the society, lacking the means to adequately feed and create jobs for its new citizens, generates high emigration, engages in wars (foreign and/or civil) and suffers from various social and natural disasters. The Irish potato famine of the mid-1800s, accompanied as it was by massive waves of emigration, is a good example.

The key to the demographic trap is family size. Women in the poorest countries tend to have large numbers of children – five or more. They generally wish to have fewer children, but they and their husbands or partners lack the knowledge and means to manage their family size. Death rates in these countries, from infant mortality on up, are high.

Research shows that the key to avoiding the demographic trap is, ironically, information – and the ability to communicate it effectively. In societies where women, in particular, are literate, and encouraged to plan their family's size, they tend to do so. Literate mothers tend to keep their children, of either gender, in school, increasing their literacy rates and economic potential and decreasing their likelihood of having children of their own at an early age.

Religion need not be a barrier to success in reducing birthrates. Strongly Catholic Italy, far from having too many children, has an unsustainably low birth rate and a declining population. Iran, a strongly and explicitly Muslim society, saw birth rates fall by roughly half in less than a decade when it instituted a family planning programme in the late 1980s.

The alternatives to literacy and family planning information are awful, especially given the many difficulties that people face in the coming decades.

In crowded countries with growing populations, uneducated young people end up on marginal lands with lack of water, excessive heat and desertification. Or in crowded cities with little chance of finding work or support. As water shortages, food shortages and other crises hit, these are the people most likely to suffer from malnutrition, sicken and die.

Or, in the wrong circumstances, to kill each other. The Rwandan genocide of the early 1990s was preceded by just this sort of overcrowding and despair. Hutu killers made sure to kill every single family member among Tutsi victims so no legitimate claimants would be left to (often tiny) family plots of land.

The demographic trap threatens to keep populations rising in some regions well beyond 2050; or, worse, to create pressures in which millions are born, only to die of environmental and social crises, quickly replaced by a large new cohort with similarly dire prospects. With few places left on Earth to emigrate to, some populations may be controlled not by improved health and welfare, leading to voluntarily lowered family sizes, but by waves of malnutrition, disease, conflict and death scything through large birth cohorts.

The current Zimbabwean crisis, ongoing wars and massacres in central Africa, and other recent conflicts have begun with demographic and environmental crises, which will only become more common in the decades ahead. Increasing population alone would be enough to cause many conflicts; when it's combined with climate change, which will make productive land marginal and marginal land useless, while suppressing productivity on what's left in random, unpredictable ways, crises may become the norm in too many places.

Scarce water and Peak Food

The direct inputs to production of grain, fruit and vegetables are water, land, sunlight, seeds and fertiliser. Most fertiliser today is artificial, made by a power-hungry process which fixes nitrogen from the atmosphere, and farm machinery plays an ever-greater role, so fossil fuels are also a critical, albeit indirect, input into food production.

Water availability is a crucial limiting factor for food production. The potential for increasing water shortages is not widely understood, and yet they are potentially threatening to the lives and livelihoods of millions, even before climate change is considered – and much more so afterward.

The first thing that's not widely understood is just how water is used. An adult only drinks a few litres of water a day. But raising food takes 1000 litres of water for each kilogram of grain. When grain is then used as animal feed, which requires up to seven kilograms of feed per kilogram of meat, the water used by food multiplies. We each, in a sense, eat thousands of litres of water a day.

Even a vegetarian or near-vegetarian diet – the most water-efficient – requires 2000 litres of water a day per person to grow and process. A European diet in which meat is an important component requires 4000 litres of water per person per day. Two all beef patties, special sauce and all the other elements of an American diet require a staggering 8000 litres of water per person per day.

Aside from food production, carrying away toilet waste, washing and cleaning with water – at home and in industry – are all inefficient. Many litres of precious drinkable water are used to wash away small amounts of waste. The contaminated water – sometimes treated, often not – then flows into streams, rivers, lakes and oceans, polluting them.

Greenhouse gas emissions are also causing another, more insidious form of pollution. Scientists have recently found that high CO_2 levels in the atmosphere are not only acidifying the seas, as described below, but some groundwater and freshwater as well. Perhaps we'll soon be pouring baking soda into our lakes and wells to protect ourselves from yet another effect of greenhouse gas emissions.

Heavy water use and water waste are multiplying as China and India, in particular, pursue the Western model of development – the "American Dream". As with fossil fuel consumption, we are moving quickly from a well-off swathe of 1 billion people, mostly located in well-watered parts of North America and Europe and using water like there's no tomorrow, to 3 billion or more, with most of the new heavy users located where water is scarcer and becoming ever more so.

Many people are worried about water shortages in various parts of the world, but few have put all the pieces of the puzzle together. At its core, the water problem is firstly a food problem, and vice versa.

The Green Revolution – not environmentalism, but the doubling of crop yields worldwide – was powered by abundant water tapped wherever productive wells could be drilled. Well water can be drawn when needed and piped where needed, making it far more efficient than traditional irrigation from rivers and streams.

Well water has allowed much of China and India to go to two crops a year per field – which couldn't be done without lots of water available just when needed – nearly doubling output. Wells are tapped using relatively inexpensive electric motors that have been available, and gradually improving, for about fifty years.

But almost no major user of well water in the world restricts withdrawals to allow their aquifers – where the water is drawn from – to recharge. India subsidises both the cost of water to farmers and the cost of electricity used to drive the pumps that deplete it. Even well-off countries like the United States, which has in recent decades done a marvellous job with soil conservation, just keep pumping and diverting water.

So today, more than half a century into the Green Revolution, the big issues for water are:

- Roughly 70% of water is used for agriculture, 20% for industry and 10% for homes
- Much of the grain grown today, especially in the developing world, is efficiently watered from wells
- Most of the wells are drawing down their local water table to ever-lower depths
- It's looking likely that an increasing number of wells in various regions will become uneconomically deep, or simply fail, beginning in the next few years – just as climate change impacts are increasing
- Climate change threatens to cut off much of the world's non-aquifer supplies by disrupting rainfall and eliminating glaciers in temperate zones, with their precious annual snowmelt.

Of the billions of people alive today, many hundreds of millions are dependent on the extra crops that are only grown by drawing down local water tables; on snowmelt and glacier melt; or both. When the aquifer becomes uneconomic or fails; the snowmelt or glacier melt goes; or both, crops will be watered only by rainfall.

The "extra" grain will disappear, leaving nothing to feed the people dependent on it. And supplies of grain grown directly from water falling as rain are less reliable, dependent on the amount and timing of rains, which become increasingly unreliable (and unhelpfully "bunched" into irregular episodes of more intense, even damagingly intense, rainfall) with climate change[5].

The monsoon rains of South Asia are of particular concern. Half the world's population – about 3 billion people – depends on crops watered by these rains. Their failure – or even significant shifts in their timing and the amount of rain delivered – threaten this irreplaceable food supply.

There are additional concerns as to arable land, driven – no pun intended – by the move of tens of millions in the developing world to a Western lifestyle:

- As car use grows, the best land for farming is often just as attractive for development, roads, parking lots and so on, taking it out of production;
- Agricultural land is needed for biofuel production, increasingly driven by rich country mandates, and for biologically produced drugs (biopharma);
- When cities and industry run short of water, as will happen more and more with climate change, they buy it from nearby farming areas, sharply lowering the land's productivity, or taking land out of production entirely;
- Overgrazing by livestock and overuse of trees drive rangeland to desert;
- Once established in an area, desertification – worsened, of course, by climate change – reduces farms to rangeland and rangeland to desert.

The bottom line for grain production is stark:
- Growing populations and increased meat production are expected to double total grain needs by 2030;
- Grain production depends on use of energy-intensive fertilisers, whose price is likely to rise in coming years;
- Grain production has dropped slightly from its all-time high in about 1990 and further decreases are on the cards;
- Annual grain output per person and world reserves of grain are dropping;

---

[5] Warmer air can hold more water vapour, so the tipping point of 100% humidity which precedes precipitation takes longer to reach, meaning less frequent rainfall. When the tipping point is reached, the greater amount of water stored in the air means that the rain can be greater in intensity.

- Grain yields are estimated to fall 10% for each rise in temperature of 1°C;
- When the thermal maximum for a crop is reached, even for a brief period, it stops producing, and slightly above that temperature, it dies;
- Food markets are global, so a shortfall anywhere means skyrocketing prices everywhere – as seen in the 2008 food price crunch, partly driven by the use of cropland (and water normally used to grow food) to grow crops for use in biofuels.

We may have already reached Peak Food – the highest production of food may be in our past. At the same time, increased meat eating and increasing population pushes the need for grain and fisheries output upward.

Much trade in grain can really be thought of as trade in water. When an area runs low in water, the region stops growing grain and those who were dependent on the output buy grain in instead. Buying a thousand kilograms of grain saves a country a million litres of water.

With meat being even more water-intensive, buying in a thousand kilograms of beef may save a region a billion litres of water. This assumes, of course, that there are other regions with sufficient extra fresh water to invest in meat production. (In addition to being water-intensive, meat is greenhouse gas-intensive as well; as mentioned above, it's been estimated that a pound of beef has 100 times the carbon footprint of a pound of carrots.)

This is the source of the advice environmentalists often give to "eat lower on the food chain". Unfortunately, people tend to respond to the iron fist of rising prices more than to the velvet glove of good advice. It may only be when a pound of beef costs 100 times more than a pound of carrots, and other meats and milk are priced by their impact as well, that consumption, and the attendant impact on climate change and water usage, drops much.

Goats, sheep, pork and particularly chicken are more efficient at converting grain to protein than cattle, so represent less investment of grain and, therefore, water. Some farmed fish such as carp are even better, operating at almost 50% efficiency – two pounds of grain for one pound of fish.

The sea seems to be a logical place to turn to if grain production suffers. But fisheries are also overtaxed; many have collapsed already, and others are in danger.

Warmer waters plus environmental damage from nitrogen runoff – excess fertiliser taken to the sea by rivers – creates ever-larger dead zones in which microscopic life explodes to consume all the nutrients and take up all the available oxygen, then itself dies, leaving a watery wasteland.

The remaining productive areas of the seas are seriously overfished. Fishing today is an increasingly high-technology and energy-intensive enterprise, with additional effort and investment yielding meagre returns. It's been reported that the stock of large fish in the oceans today is 10% of what it was half a century ago, a needless side effect of overfishing. Recent research shows that the largest catches ever recorded – Peak Fish, if you will – occurred in the late 1980s.

As if dead zones and overfishing weren't problem enough, as increasing CO_2 levels in the atmosphere cause increased CO_2 uptake by the oceans, they become more acid. This occurs most strongly in sunlit surface waters – where most marine organisms live. This is a well-known problem that has recently been found to be worsening faster than scientists had predicted.

A strong acid like battery acid has a pH of 0; a neutral solution like pure water has a pH of 7; and a strong base (the opposite of acid), such as bleach, has a pH of 14. The acidity of seawater before industrial pollution was 8.2 on average, which is somewhat basic.

Average seawater acidity has now fallen to pH 8.1, which seems like a small change, but is 30% more acidic than before. It's now thought that small sea life will begin to be killed, killing corals and causing disruption to all marine food chains (and therefore many of humanity's food chains) when greenhouse gases reach a level of 450ppm of CO_2 and equivalents, which will cause seawater acidity to fall to about 7.9 – more than double pre-industrial acidity. Coral reefs alone are home to fish that contribute to the diets of a billion people.

(The problem of acidity in the seas is also important because various schemes to cool the earth – to compensate for global warming's heat without the need to reduce emissions or remove CO_2 from the environment – fail to deal with the ocean acidity problem.)

Such losses will be a huge shock to humanity, as it increasingly depends on the sea for food and feed. Livestock and farmed fish are increasingly fed on fish meal from the seas, so it's not only the direct effect on wild fish that we have to fear.

With grain production challenged by reductions in fresh water availability and prices rising due to shortages and the impact of Peak Oil, and with fisheries under multiple threats, the bottom line is that we are unlikely to be able to feed the additional 70 million people born each year, mostly in poorer countries.

These same countries are seeing the emergence of Western lifestyles, with millions moving to high-consumption lifestyles, including meat consumption, that further pressurise resources. The competition for resources can create not only failures and famine, but rapid, worldwide price rises. Big disruptions are likely as emergency efforts are needed to feed growing numbers of people, with accompanying failures that leave many to starve.

This is an extremely brief and high-level description of a very complex and alarming situation. It deserves a book of its own – yet it's still less serious than the core momentum of climate change. Climate change is part of the food and water problems – and humanity must address it in order to have a decent chance at addressing the other problems.

Air pollution and climate change

Air pollution is a great scourge of humanity, costing many millions of people their health and even their lives. It particularly affects infants, children and the elderly, its effects hidden among the many health problems it causes or contributes to. Many billions of dollars are spent every year to reduce air pollution and to deal with its many ill effects.

The US and Europe suffered greatly until clean air laws were passed in the 1970s, with London and Los Angeles only the most famously smoggy – in London, the euphemism used was "foggy" – of the many cities affected.

Asia suffers even more today, with smog from many cities merging into atmospheric brown clouds – ABCs – which sometimes cover a multinational arc around populated areas in a single vast expanse of dead air. The amount of sunlight reaching the ground under an ABC is often reduced by 20% or more.

Ironically, this scourge may be partly saving us from an even worse one – climate change. It's believed that the West's smog of the 1960s and 1970s, and the East's ABCs of today, may be responsible for reductions of perhaps 0.5°C in temperatures when at a peak. It may not be an accident that average global temperatures rose from 1910 to mid-century, paused during the West's smoggy decades, and resumed rising steadily with the US Clean Air Act and equivalents in other industrialised countries – yet today, with ABCs prevalent, are again somewhat lower than many models predict they should be.

Yet pollution – not only of the air but of water and soil – is already a huge, and may become an immense problem in the next few decades. China and India are only getting started with their moves to Western lifestyles, and are experiencing rapid population growth that will only worsen their problems, yet are already poisoning all too many of their citizens.

Though warming may be reduced by the "cloud cover" effect of smog, warming's effects are accelerated by soot. As mentioned in Chapter 6, soot particles falling on ice- and snow-covered areas act as little bullets of heat, melting the snow around them. This has an immediate effect of reducing the reflectivity of existing snow cover and will have a greater effect as areas see greatly reduced snow and ice cover, reducing their reflectivity.

Pollution also damages agricultural productivity, reducing the amount of sunlight reaching plants, sickening them and reducing their growth, all while soot attacks the icy sources of the water they need during growing season.

Air pollution in particular, with its accompanying soot, will simply have to be reduced, yet the same economic and demographic pressures that make it hard for developing countries to combat greenhouse gas emissions will make it hard to combat air pollution as well.

In fact, turning our view around, the willingness of Asian countries to endure ABCs, and other effects of pollution, beyond the worst the West ever experienced, bodes ill for the likelihood that they'll invest to avoid the seemingly less direct impacts of climate change. (Just as the long-term tolerance of the West for many kinds of pollution has been a bad omen for its ability to protect people or nature from climate change impacts.)

The rest of the world may benefit from a modest reduction in warming – and from a vital clue in how particulates, minus the toxicity and contribution to melting, could be used to battle climate change – even as large parts of the East suffer serious or even catastrophic consequences from pollution and the resulting reduced water availability.

Extinctions

Climate change has already helped exacerbate other factors in leading to great losses in the natural environment, with extinctions spiking upward. The combination of climate change, water shortages, the response to food shortages, deforestation and increased pressure on the environment due to climate change will greatly increase damage to the natural world in the coming decades, with massive extinctions only the final, sad result.

This topic could easily fill a book of its own. I haven't spent much time on it here because, if we can't get motivated to save ourselves, talking about losses in the natural world – no matter how tragic and permanent – won't add much.

But we have never properly studied, measured or accounted for the services provided to humanity by the natural world. The long list includes oxygenating the air, cleaning the air, fertilizing crops, cleaning water, providing additional sources of fuel and food and much more. As we start to lose them, we'll feel the losses deeply. Not having studied the topic properly before, we'll only find out the hard way – by losing them – just what these services are and just how valuable they are (or were) to us.

The irretrievable loss of most of what we know as the natural world may also cause, appropriately, a lot of grief, and more or less panicked efforts to save some of the remaining bits and pieces before it's too late. In hindsight, a massive and coherent response was needed beginning in, perhaps, 1962, when the famous book *Silent Spring* by Rachel Carson came out. If we had taken solid steps to save the wild environment beginning then, we would probably have laid the groundwork to save ourselves from climate change now as well.

The opposite is true today. The remaining wild world can be seen as a shield for humanity, with massive efforts to save it as the cutting edge of efforts to save ourselves. The change in consciousness required to take on such a cause would serve humanity well.

The direct effects of climate change

In the years from today to 2050, the direct impacts of climate change will become greater and in some cases more dramatic, but will also express themselves significantly as a worsening of the problems noted above, in particular lack of water and lack of food.

Even if emissions were to start being cut tomorrow – in itself unlikely – warming directly due to emissions of at least an additional 1.3°C, for a total warming over pre-industrial levels of 2°C, can be expected by 2050. In fact, keeping warming to no higher than this level – in the, to my view mistaken, belief that such will prevent runaway climate change – is the main goal of current emissions control efforts.

The main benefits of any cutbacks will come in later decades and centuries. As shown in the previous chapter, carbon sinks in nature will also be tipping, adding to warming, with an impact of a full degree or more by 2050 likely. While the larger impacts will manifest themselves in the second half of the century, a total impact of 3°C by 2050, plus the financial, social and psychological impact of knowing that much worse is on the way, will be more than enough to create huge strains.

The direct impacts of higher temperatures in the next few decades alone will be substantial. An increase of 3°C means an increase over continental land masses of perhaps 5°C (about 9°F) in today's temperate zones – even more than that towards the poles. This will shift both natural habitats and crop-growing zones rapidly toward the poles, disrupting both.

Worse, adaptation will be an ongoing and increasing strain as change continues. And even worse, the weather, while changing as the underlying climate zones shift, will also become more erratic and less dependable. The erratic weather will be accompanied by higher sea levels, increased flooding and a greater number of intense storms.

The "noise" of erratic weather will frequently obscure the "signal" of ongoing shifts in climactic zones, making planning and investment extremely difficult, reducing total food production and making annual harvests ever more uncertain.

As mentioned in previous chapters, the IPCC's 2007 prediction of a half-metre rise in sea levels by 2100 ignored effects from glacial melting; credible predictions are now being made of a rise of half a metre – perhaps even a full metre, according to one expert – *by midcentury*. Each incremental rise in sea levels strengthens the effects of (more and more frequent) intense storms; the full rise will threaten the habitability of, for instance, coastal areas of Norfolk and most of London.

As with carbon sinks, there may be non-linear – a scientific term meaning, in this case, "sudden and frighteningly rapid" – changes in sea levels as well. Research has shown sea level rises of several meters per century in the geologic past; we are inviting that kind of sea level rise again, with the key question for us today being when it begins. The Wilkins Ice Shelf in Antarctica and the Ayles Ice Shelf in northern Canada are two recent, dramatic examples of completely unexpected collapses. If nature continues to surprise us in this way, even seemingly aggressive current estimates of sea level rises may be, well, all wet.

The human response to climate change

The human response to climate change will cause its own problems – as first-generation biofuels and disputes over emissions have already done. (James Lovelock has said that the most effective single action humanity could take against climate change is, ironically, to stop producing ethanol.) Global responses are needed, but humanity is organised nationally, and even by smaller units – republics, states, provinces, cities, regions and ethnic homelands within and sometimes crossing national boundaries.

The response to climate change is likely to be in fits and starts, a sometimes chaotic melange of individual decisions, business and regulatory actions and government mandate. It might be possible to farm in a given area, but become impossible to get a loan or insurance to support investment for farming there; a single incident may largely empty an otherwise safe area of people, while a truly imperilled area nearby remains crowded until disaster strikes.

There is no model for successful response to a steadily escalating threat with consequences that are so extensive, yet almost impossible to predict in detail – and with any given region moving back and forth across the continuum between relative safety, increasing discomfort and actual peril. Individuals, businesses and governments, who tend to frame their choices in terms of "doing something" or not, will wait too long, suffer, then react abruptly or overreact, again and again.

Many books can and should be written about how societies should adapt to the already certain impacts that we'll face from climate change, in addition to the impacts we'll face if we don't stop deforestation, stop emissions and reverse runaway climate change.

But one of the most interesting, for any sociologists who remain in a century or two, and most potentially troublesome impacts, is the range of human responses as people gradually realise just how bad things are going to get. The full grief cycle may come into effect – shock, denial, anger, bargaining, depression, testing and acceptance, but repeated over and over for different levels of understanding and in response to different specific impacts.

People will experience – or struggle to stave off experiencing – tremendous levels of guilt and, the other side of the same coin, a tremendous desire to assign blame as the worsening of prospects for themselves and future generations becomes clearer and more difficult to avoid.

The UK's Sir John Beddington, chief science adviser to the UK Government, has recently spoken out to highlight 2030 as the year by which these interrelated problems – water shortages, lack of food, extinctions and the direct effects of climate change – must all be at least well on the way to being resolved, or truly dire consequences for all humanity will ensue.

This urgency is part of the reason why it's so important to understand as soon as possible – now – that we're already in runaway climate change, which Beddington did not take into account. The longer the realisation is put off, the worst off we will be in fighting the consequences, and at the same time, the harder it will be to deal with the emotional response people will have when the new realities finally become inescapable.

Toward Hot Earth?

Humanity has a huge challenge in the coming decades - its greatest ever. We must do the difficult, transformational work needed to head off runaway climate change while also coping with rapid population growth and the climate, food and broader environmental impacts that will make it hard to support the world's current, let alone its projected population.

Current trends in emissions and tipping carbon sinks lead us to a world roughly 10°C hotter than pre-warming temperatures in this century. Even considerable success in fighting these trends may lead only to postponing this fate for some decades, rather than retaining the world we have. Only a combination of alertness, hard work and luck will allow real victory - a return to the "green and pleasant land" that we had on much of the Earth, as well as richness and diversity in the seas.

But what if the effort fails - what if runaway climate change continues?

Then we will move to a changed world that I call Hot Earth. Our planet would be more than 10°C hotter, with deserts expanded to cover roughly one-third of the world's land area. Most of the species that make up today's natural world would be dead, killed on land by the direct effects of heat and by ecosystem collapse; in the sea by a combination of heat, acidity, anoxia and ecosystem collapse.

The characteristics of Hot Earth can be summed up in the acronym HIDE:

- **Hot**. 10°C hotter global average temperatures, with much greater increases at the poles. Temperatures would not only be hotter but much more even planet-wide, greatly changing all sea and air currents and upending weather patterns worldwide; the specific results in a given region are unpredictable.
- **Ice-free**. The hotter Earth would be free of permanent ice, therefore far less reflective - and more absorbent of heat than today. Precipitation would be retained neither as snow and ice nor by vegetation, but would tend to flow straight through to the seas - which would rise as the ice finished melting to a level more than 120 metres higher than today.
- **De-evolved**. With change happening so fast, the biosphere would be caught between two stools. The last few million years of evolution would be rolled back, with most higher life forms on land and sea nearly or completely wiped out. (A last-gasp effort by people might transplant some plants and animals to new regions, much farther from the equator,

where they could be sustained going forward.) But a multi-million year wait would lie ahead before new life forms could evolve to appropriately populate the new environment.
- **Empty**. In comparison to today, or perhaps to anytime since shortly after Snowball Earth conditions ended 600 million years ago, the Earth will seem nearly bare. A third of the land area of Earth or more will be outright desert. Far less vegetation will cover most areas, and what is left will be simpler, supporting a much-reduced stock of animals and insects. Most fish will be gone, with jellyfish perhaps the most complex life able to thrive in acidic seas.

We don't know what the "carrying capacity" for humanity of Hot Earth would be. It would depend tremendously - perhaps completely - on the preparations people would be able to make. If people have been fighting one another to retain as much as they can of the Earth's current bounty right through the onset of Hot Earth, its conditions might then kill off most of the survivors; if people have been planning and preparing, a population of a billion or even a few billion might be sustained.

This is part of the reason it's so important that we not only begin to fight, but to understand runaway climate change immediately - so we can understand what our options are going forward, and just what we might have to prepare for.

But a situation in which a good result is a human population 50% less than today's is clearly unacceptable. We owe it to ourselves to take the many difficult steps needed to retain and even replenish our current environment, so Hot Earth does not become a reality.

Chapter 9. Fighting Climate Change

> **In this chapter**
> Steps for fighting climate change:
> 1. Set out principles
> 2. Get the science right
> 3. Stop digging
> 4. Zero tolerance 1: power
> 5. Zero tolerance 2: transport
> 6. Start removing carbon
> 7. Lower the temperature

"There is nothing more difficult to carry out, nor more doubtful of success, nor more dangerous to handle, than to initiate a new order of things. For the reformer has enemies in all those who profit by the old order, and only lukewarm defenders in those who would profit by the new order, this lukewarmness arising partly from fear of their adversaries, who have the laws in their favour; and partly from the incredulity of mankind, who do not truly believe in anything new until they have had actual experience of it." Niccolò Machiavelli

The Machiavelli quote above is of course favoured by all who attempt new things. There are two "old orders" threatened by my assertion of runaway climate change in this book: those who have a vested interest in the processes that led us into climate change; and those who have a vested interest in a specific diagnosis and set of recommendations less sweeping than that presented here.

But there's an additional sting in the tail of the quote – that mankind "do not truly believe in anything new until they have had actual experience of it." In the case of climate change, this is a tough standard, as extreme weather, food shortages etc. lag by decades the trends in greenhouse gas emissions that cause them. They hit haphazardly, and they don't come stamped "climate change". Machiavelli's insight helps explain why readiness to act on climate change has been so slow to gel – and why "old orders" standing against action find it so easy to maintain support.

People's understanding of, and response to climate change can be divided into three main schools:
- A "know nothing" group not particularly interested in whether climate change is really happening, or somehow convinced that it isn't, despite all evidence to the contrary. The true "deniers" in this group will actively undermine research into, and scientific findings about, climate change, as occurred with the Bush-Cheney administration and efforts by a lobbying group led by ExxonMobil to obscure the issue.
- The group that holds to the 2007 Consensus, though there's a continuum from optimists through to pessimists in this group. This is the vast majority of informed opinion, since they draw their information from the only really authoritative, though deeply flawed, source, the IPCC and its 2007 Report.
- Last is a small and unorganised group that knows or strongly suspects that the 2007 Consensus is too optimistic, but that lacks a flag to rally around. Increasingly, scientific findings, arguments like the one presented in this book and real-world events will peel off adherents of the 2007 Consensus into a new grouping that comes to understand that runaway climate change has already begun – and that it needs to be fought as such.

The first step in the battle against runaway climate change is to create a new home for these lost souls, based on the latest information and a less optimistic reading of the received wisdom. I'll provisionally call this new pole of opinion the Runaway Consensus.

How long might it take the Runaway Consensus to become mainstream? The pessimistic view is that it will take either the current adult generation shuffling off this mortal coil – another two to three generations, about 50 years – or a series of disasters so great that no one can ignore them, what Joe Romm of the Climate Progress blog calls "multiple climate Pearl Harbors", forecast to occur over the next few decades.

While this view is highly plausible, we can't wait that long, so let's hope for a quicker change. The world is hugely dependent on the IPCC process for its information about climate change – and has been, as I described in Chapter 3, poorly served by that dependence. Our best hope is that the scientists on the IPCC wake up, to their own mistaken promulgation of their baseless Scenarios and to the political manipulation of scientific results. And that they then force the policymakers to agree on a full Report that reflects current reality and reasonable projections from it, or prepare a minority report unedited by policymakers.

The soonest this could happen is in the next IPCC Report, scheduled for 2014. Or it could take until the one after, in 2020 or thereabouts. Either may be the IPCC's last chance; the IPCC risks losing all credibility if its reports fail to reflect an increasingly large body of evidence, and stay so far outside what will become the emerging scientific mainstream, for very many more years. But the world will be better served by a revived IPCC process than by any other scenario. (No pun intended.)

Too late by half?

The Runaway Consensus has a question at its core: is it already too late? Are we doomed to face perhaps 10°C of warming in this century, and the relatively quick collapse – over the coming decades and few centuries – of all carbon sinks?

The short answer is "yes". With anything resembling current levels of effort and focus, and with current technology, we're doomed. We do have the technology to cut emissions relatively quickly, if doing so is treated as a true global emergency; but we – and "we" would have to include every major country on Earth – don't currently have the will to deploy it faster than the two generations or so such changeovers. And cutting emissions on its own is no longer enough.

The only thing certain to generate the will needed to move faster is Romm's "multiple climate Pearl Harbors" – but those are probably still some years off, meaning more emissions and more carbon sink damage in the meantime. And, as with the original, each of the "Pearl Harbors" will enfeeble the efforts humanity might then be prepared to make in response.

We also lack the technology to easily, cheaply and reliably remove carbon from the atmosphere and oceans with minimal side effects, as needed to halt runaway climate change in its tracks, as well as the political structures to generate international agreement and financial support for doing so.

However, as George Monbiot has bluntly put it, we can't afford to give up, so we won't. Warming of 10°C this century is far beyond being unacceptable. Whenever the realisation hits that only an intense and focussed response can possibly forestall disaster, the needed response is likely to coalesce. Eventually, all of us – even today's sceptics and deniers – will battle it with everything we can bring to the battle. We will have to assume that we will then win.

The Runaway Consensus needs to set out a framework for solving the problem, then generate the political will to act on it – helped by increasing scientific and policymaker support on the one hand and the threat or reality of climate Pearl Harbors on the other. The second factor reinforces the first; politicians have to keep getting campaign donations, and keep getting elected, which encourages dawdling on climate change; but neither does any of them want to pip George W. Bush for the title that history will assign as the Neville Chamberlain of climate change – with John Howard currently in the lead for the "Best" Supporting (non)-Actor gong.

A framework for saving the Earth

In Alice in Wonderland, the White Queen says she has sometimes believed as many as six impossible things before breakfast. Perhaps the right approach for addressing runaway climate change is to believe – and work toward – two (nearly) impossible things.

The first (near-)impossibility is that we embrace and extend the emissions reduction framework set out by the 2007 Consensus. Each country must adopt – and do all in its power to ensure that other countries adopt – the widely agreed emission reduction goals of 20/20/20, the emerging goal of 80/80/50 and, for completeness, a new goal of 100/100/70.

That's 20% of energy generated from renewables, and a 20% global reduction in emissions, by 2020; 80% renewable energy and 80% reduction in emissions by 2050; and, aided by the gradual decline in world population expected to begin in 2050, 100% renewable energy and zero emissions by 2070.

These goals are probably achievable for the developed world, as originally prescribed by the 2007 Consensus, which intended to let developing world emissions grow moderately in the meantime. However, we need a tougher ultimate target than the 2007 Consensus – a 100% reduction by 2070 – and we have far higher developing world emissions and a far higher trend than previously estimated. So it's the world as a whole that needs to meet 20/20/20, 80/80/50 and 100/100/70.

The only workable solution that includes the whole world is based on contraction and convergence, as described in the next section – more rapid cuts in the developed world, even beyond, for instance, 20/20/20, combined with a rapid slowing of the rate of emissions growth in the developing world. As emissions per person for developed and developing converged with each other, then it would all be contraction, toward zero emissions per person in 2070.

There isn't an explicit goal from the 2007 Consensus for deforestation. However, given the multiple kinds of damage it causes – contributing to emissions directly, reducing the uptake of emissions from other sources, enabling damaging development such as new roads and stockyards, and further infringing on the tattered natural world – deforestation has to stop immediately. Achieving this will take a worldwide financial mechanism to protect the world's remaining forests, with reforestation then used to help meet other goals listed above and below. The new financial mechanism could then be extended to the technology sharing and implementation financing needed to meet the ever-more-restrictive emissions goals.

Achieving these ridiculously strict goals on the emissions and deforestation fronts might just provide a basis for achieving the second (nearly) impossible thing: directly taking on runaway climate change. Aided by steadily declining emissions and, no less importantly, by the new-found international cooperation making these cuts possible, the world's scientists and engineers – already busy reducing emissions – would be tasked with creating a geo-engineering programme to offset the effects of collapsing carbon sinks. Some of the current ideas about how to do this – painting surfaces white; floating white, biodegradable stuff on the seas; putting sulphur particles in the upper atmosphere for reflectivity; chemically neutralising the seas' increasing acidity – have been mentioned in this book. There are other ideas, and there would need to be still more.

A much-improved version of the model shown in Chapter 7 would become even more important – not as a road map to Doomsday, as it might seem at present, but as a diagnostic tool for identifying the points that will otherwise be attacked by, and the weak spots of, the enemy, runaway climate change.

This approach takes the 2007 Consensus at its word on emissions and gins up a parallel effort to deal with collapsing carbon sinks. These paired emissions reduction and geo-engineering efforts would become the overall framework in which the steps described in this chapter would take place.

It might turn out that a somewhat different mix between emissions reductions and geo-engineering solutions will be optimal, but it's doubtful that any such change can be too great; since each of the two overall tasks is nearly impossible, neither can bear the weight of a significant failure to achieve its goals by the other.

Even if such an overarching effort were to fail, we would know the enemy much better; we would have gained many decades of valuable time; and we would know the effectiveness of, and the limits of, our own capabilities, both technological and in our ability to cooperate in the face of the most truly global problem mankind has ever confronted. Our odds of saving much of what we have already achieved to date, as well as what we additionally would by that point have achieved, would be significant, even on a planet moving inexorably toward Hot Earth.

If the effort did fail – if we couldn't achieve the emissions reductions and geo-engineering successes needed to avert runaway climate change – we would then need to embark on a wholesale programme of adaptation to a world 10°C hotter than today. Buying time for adjustments in how food is grown, and for populations and industry to make an orderly move away from doomed coastlines, are just two examples of the benefits of changes that buy even a few decades of time.

The goals I think appropriate for reducing global, not just developed world, emissions, and for geo-engineering – what I call the Runaway Consensus – are:

1. Set out strong founding principles
2. Get the science right
3. Stop digging
4. Mass-produce carbon-free power
5. Move to carbon-free transport
6. Start removing carbon
7. Lower the temperature.

Goal 1. Set out strong founding principles

I believe that part of the reason that the 2007 Consensus was so unwontedly optimistic – even the 2007 IPCC Report, supposed to have been grounded solidly in science and its fearless pursuit of the truth – was fear itself. Fear of the truth that might emerge if one pursued scientific findings to their logical conclusion, fear of rejection by the public and policymakers, and fear of derision as a naysayer or doom-monger. (Negative forecasts on climate change are still sometimes called "climate porn" in the press.)

It's been widely believed that the public has to be brought along gently; those who press for a completely accurate assessment have been called "defeatists" by those who have one specific, but insufficient programme or another mapped out against climate change.

Al Gore, for example, has described how he sometimes has trouble staying optimistic. But he's trusted for his brain as well as his motivational skills, so perhaps he shouldn't try to be optimistic, but just call it as he sees it.

So the first principle in a new consensus on climate change needs to be honesty, including honest answers to the questions raised by this book and other sources.

The next principle needed is fairness. There's much discussion, for instance, of how to allocate goals and quotas for various tax and market-based approaches.

The public might be ahead of leaders here. A 2008 poll asked about "extra" emissions allowances for developing nations. People in developed countries were ready to sacrifice some of their own allocations to allow developing countries to grow; but people in developing countries preferred cuts in their emissions, even from a low base, to extra allowances for themselves. These results may shift when translated into specific choices, but the underlying mindset is promising.

With greenhouse gas emissions in one place quickly affecting the level everyplace, the UK's Aubrey Meyer hit on, and has actively promoted, the simplest and only fair approach: contraction and convergence, sometimes called C&C. Every person on Earth should have the same initial allocation for CO_2 emissions as every other. Nations will then have allocations based solely on their population. And countries can then share out their allocations among industries and individuals as they see fit, and buy allocations from one another where needed.

However, even Meyer, visionary that he is, still sees room for a small climate change allowance per person that, in aggregate, leaves substantial global greenhouse gas emissions per year.

It now seems that the Earth's climate, in the pleasant (to us) state we have experienced from the end of the last Ice Age 11,000 years ago until very recently, was quite delicate. We know it was delicate on the down side because temperatures have repeatedly, over the last 650,000 years, slid down from the levels we were experiencing before warming into Ice Ages. With our emissions and the resulting warming, we've found the climate was delicate on the up side as well. The current warming trend began in 1910, when CO_2 levels were only about 300ppm – less than 10% higher than when human emissions began to increase, around 1760.

This shows that excess greenhouse gases are, in the context of our desired (pre-warming) climate, quite a dangerous pollutant. (As President, Barack Obama has presciently and repeatedly referred to "carbon pollution" in his speeches, laying the political groundwork for regulating it.) They're not dangerous in a short timeframe; but in the context of maintaining, in the decades and centuries to come, the specific, post-Ice Age climate, better environmental health and sea levels that we had before warming began, they're bad actors.

Another way to look at the same situation is that we're running our food production, energy and other systems – all of which include both natural and human components – at or beyond capacity to support 6.5 billion people. Yet we're about to add many more people, and on track to make billions of people better-off (and therefore much more resource-intensive to support). We can't afford to have the underpinnings of all our systems – our lovely, pre-warming climate and natural environment – collapse now.

When you're dealing with a dangerous pollutant – such as lead, radioactivity or CFCs – you don't begin with how much you'd like to emit, then adjust downward to some kind of compromise level. You start at zero, then allow a very small margin of emissions if this is shown to be harmless. (Both lead and radioactivity occur in nature, so not exceeding natural levels of exposure is a reasonable target for these pollutants; a similar approach may be appropriate for greenhouse gases.)

To use a very crass example, it's as if you have a dog that poos on the living room carpet. The very first thing you do is get rid of the poo. Then, you don't seek to reduce the frequency of pooing on the living room carpet by some modest annual percentage. You don't make a mark on the baseboards as a poo target, then make a half-hearted effort to keep the accumulation of poo below the limit. Your living room is important to you. So you quickly train the dog so he or she never, ever does it again. If the goal can't be achieved rapidly, you get rid of the dog.

So it is with greenhouse gas emissions and deforestation. To maintain our favoured environment and sea levels, we need to reduce greenhouse gas emissions and deforestation to, effectively, zero. Then we need to remove the greenhouse gas emissions already in the atmosphere and water. And we need to cool the planet to counteract the warming that has already taken place and that will continue until emissions and deforestation are stopped.

For deforestation, we not only need to stop cutting down and paving over greenery, we need to seek to restore the levels of greenery found before human activity started whacking away at it, and restore accompanying wildlife in the land and seas.

Zero, though dramatic, is a number people understand. This zero-based approach will eliminate distortions we see today. High-emission countries – for instance, those with lots of coal-fired power plants and/or heavy industry – demand, and get, extra allowances for emissions, while continuing to pollute heavily today and for many years to come. Instead, high-polluting countries need to be moving fastest to zero emissions, and paying for the clean-up of existing greenhouse gas pollution in proportion to past emissions.

Today, low-emission countries, with few places to cut, are still expected to do so. A zero-based approach rewards today's low-emitting countries for keeping emissions low; they need make smaller cuts and little contribution to cleanup. Only this simple, fair approach will receive ongoing public support as tough choices demand sacrifice across the board.

So, to Aubrey Meyer's "contraction and convergence", I would add two words: "to zero". "Contraction and convergence to zero." To use a still-popular term, "zero tolerance" for emissions.

We also need to remove the excess CO_2 in the atmosphere and seas so the environment can recover; this should be taken back to pre-industrial levels. "280ppm or bust" might not sound very exciting as a rallying cry, but it is likely to be a necessary one. An accompanying rallying cry of "pH 8.2 or bust" – restoring the acidity of the seas to pre-warming levels, from its current pH of 8.1 and falling fast – is even more obscure, but also needed.

Another underlying principle I recommend goes to priorities. The natural environment is threatened by climate change as never before. Historic treasures are threatened by pollution and rising seas. Social and political arrangements of long standing are being put at risk by, for instance, water shortages.

The first and foremost priority – using the word in its strict sense, "that which you do first" – must, I argue, be saving lives. This seems obvious, but today low gas prices for Americans, for instance, are prioritised above saving lives in the developing world. (And, for that matter, in New Orleans.)

Too little emphasis is put today on humanitarian relief, with Western governments in particular making commitments, then not keeping them. We have many other needs in this crisis, including preserving nature, maintaining our freedoms and reducing poverty. But saving lives is the only top priority that can gain universal support and that won't fall by the wayside as extreme pressures develop. So it must be put first.

While lives must be saved today and in the future, sustaining all of our lives means sustaining what's left of the natural world. So rapid response to crises to save lives must be paired with preservation and restoration of natural environments.

The final principle is one I describe as SQA, for status quo ante. In any effort to battle climate change in general, and runaway climate change in particular, the question arises: where to stop? What's our goal for the planet – and therefore for ourselves and the natural world we're still part of?

I think the only defensible answer is: back to pre-warming temperatures, levels of emissions, and levels of greenhouse gases in the air and seas. Back to pre-rising sea levels and back to the environmental wealth and health we had before warming began.

Human populations are located according to climates – and sea levels – that have been in place for many centuries. We are essentially sacrificing the prospects, if not the very lives, of hundreds of millions or billions if we allow temperatures to rise further and to stay above pre-warming levels.

Also, it's become obvious that even low levels of additional greenhouse gases cause warming, as current warming and visible melting of glaciers began around 1910. So any "safety first" argument relating to climate change argues for a reduction to pre-warming greenhouse gas levels and to a pre-warming climate.

The same applies to Mother Nature's stock of animal and plant life. We need to do our best to restore the environment in place before industrialisation, when human populations were less than 1 billion.

What is little realized today is just how rich the natural world was before human impacts increased with population growth. Rivers once ran full of fish, and so did the seas. Great flocks of birds filled the skies and huge herds of animals roamed the land. We should seek to restore this richness, which is only possible with a return to pre-warming conditions and careful management of people's impact on the planet. Only in this way can we safely maintain human as well as animal populations.

Finally, to restore balance in the environment, there will be a need to restore ice cover at the North Pole and to see temperate-zone glaciers restored in extent and thickness. While temperature can potentially be decreased to allow this by using various kinds of geo-engineering, only pre-warming greenhouse gas levels put such goals in range.

To sum up, the principles I'm recommending are:
- Saving lives as first priority
- Honesty about climate change impacts
- Fairness and effectiveness through a zero-based carbon budget; "contraction and convergence to zero"; "280ppm or bust"; "pH 8.2 or bust"
- Return to the pre-warming climate, levels of greenery and wildlife; "status quo ante".

Goal 2. Get the science right

The biggest problem we have in fighting climate change today is that there are gaps and errors in the science, and in its communication to and understanding by opinion leaders and the public at large. And the main reason these are wrong is the two-tiered IPCC process and its hijacking by a succession of government representatives who, without enough protest from scientists, have softened the reporting of many scientific conclusions and obfuscated much of the rest.

Along with the other sins mentioned previously – narrowly based projections of sea level rises, dependence on baseless Scenarios – conclusions about the extent of climate change were deliberately downgraded in the 2007 IPCC Report. And the language used – "very likely", "highly likely" and so on – is a code, using the cautious language of science, that's rarely cracked by those outside the process.

Worse than the acquiescence of so many at the time of the 2007 IPCC Report was their lack of protest afterward. It's regrettable, but perhaps understandable, that key players might be confused or cowed at the tense ending of a difficult process; but for them to have largely stayed silent to today is difficult to comprehend. Perhaps a Nobel Peace Prize medal fogs the mind – or weakens the will.

A few scientists did protest, and their conclusions – before the "editing" process – were leaked, to some attention in the media. But the published conclusions were allowed to stand and accepted as legitimate. The more aware among the conservative press – those in the "know-nothing" camp I described above – even heralded the relatively weak conclusions of the 2007 Report as a major victory in the fight against what they describe as the conspiracy to put climate change theory over on an unsuspecting public.

The 2007 Report was a huge lost opportunity. If the IPCC process had delivered a more straightforward, accurate result, runaway climate change as a currently ongoing process might now be a topic of mainstream scientific investigation, not – at this writing – a fringe theory.

So what needs to happen now? In a word, pressure. IPCC members are representing the interests of the people of the world in the most important challenge we face, as well as the reputation of science itself as the tool humanity deploys in the unbiased pursuit of truth.

People interested in climate change should use the Internet to track, comment on and publicise every step of the process. The major inputs are scientific papers which can be either directly read and understood by, or explained to, anyone interested. The path of each major scientific conclusion or open issue needs to be traced from the scientist(s) it originates with to its appearance, or not, in the Summary of the upcoming 2014 Report.

Within the IPCC process, transparency is needed. Discussions and interim conclusions need to be made freely available. (It's ultimately taxpayers' money paying for most of this.) Any claims as to the advantages of having some discussions outside the glare of public attention have been put to rest with the 2007 debacle. Scientists will be able to be more honest and courageous, not less so, if the whole world is watching.

It's actually very late to start. Long processes like the IPCC's mostly depend on work published at least two years before the reporting year. Before that, it takes about two years for important scientific articles to be submitted, go through peer review and be published. That's after a year or so to write up the results.

So the IPCC's work, in the normal course of events, is based on data gathered five or more years before the report date. It also re-uses constructs from previous years – though one can hope the useless 2000 SRES scenarios will be abandoned forthwith.

But the IPCC is particularly resistant to new information, right from the top. The chair of the IPCC, Rajendra Pachauri, recently said: "I refuse to accept that a few papers are in any way going to influence the long-term projections the IPCC has come up with." While it's true that new information needs to be handled carefully, rejecting it a priori is hardly a scientific attitude. Most of Albert Einstein's reputation, just for one example, rests on "a few papers" – yet those few papers changed the world.

But the normal slow process of research and reporting, plus the defiant attitude at the top, means that the foundations on which the 2014 Report will be built may already be locked into place. In relation to a fast-changing phenomenon like climate change, this is too slow. An annex or update should be included and related to trends in the main report; this would also avoid both the perception and the reality that potentially alarming newer data is swept under the rug.

The IPCC should also open up to greater public scrutiny the process by which the work it depends on is proposed, funded and carried forward. This might bring in additional funding for work that otherwise would be done slowly or not at all.

This might be seen to risk politicizing science, but, within the IPCC process, science explicitly *is* politicised. (It's an "Inter*governmental* Panel", remember, with policymakers having the last word.) The only question here is whether it can be politicised in favour of humanity's broad interests, rather than only in favour of a few obstructionists working in the shadows, out of public view.

Along with a larger role in the IPCC process, the public must have every opportunity to be informed about, and participate in decisions about and action on climate change. We are in the fortunate position of still having the focus and power made available by "old media" such as broadcast TV networks and newspapers, plus growing and fast-changing new media outlets as well. If there's a media shakeout on the cards, as seems to be occurring today, which outlets best inform and engage people on climate change-related topics may help determine the winners and losers.

Getting the science right also means addressing the issue of runaway climate change. We need to have real scientists characterise carbon sinks – to come up with much better estimates than my own, herein, as to how sensitive they are to tipping, or how far they've already tipped, and how fast they're likely to contribute to warming at various temperatures. Then put the pieces together to give us a much better picture of just how much trouble we're in, and whether there are any weak spots in the expected progress of runaway climate change where we might take a stand against it.

Goal 3. Dig slower

There's an old saying that, when you're in a hole, the first thing to do is to stop digging.

Unfortunately this is, in an absolute sense, nearly impossible with climate change. Greenhouse gas emissions and the effects of deforestation are so great, and CO_2 emissions stay in the air so long, that reductions in emissions are only reductions in the speed with which greenhouse gases accumulate further. The only practical way we have today to actually remove greenhouse gasses from the air is reforestation, and it's slow, difficult, and small in effect compared to today's deforestation and emissions.

We can't stop digging right away, so we can only slow down as fast as possible. There are some contributors to climate change that are particularly damaging that need to be addressed particularly fast.

Coal burning is the biggest and most important target. Coal burning is the source for up to a third of all greenhouse gas emissions worldwide – much more than its share of power generated.

Almost as bad, coal is very sooty. This probably has some benefit in slowing warming by contributing to the clouds of pollution that substantially block sunlight at times in Asia. But the soot, landing on snow and ice, melts it quickly, contributing to the disappearance of crucial Himalayan and other glaciers, threatening the very basis of agriculture and therefore of human life in India, China, Vietnam and elsewhere.

So reducing coal use seems like a natural for sharp reductions in climate change and its impacts. But coal use is actually increasing steadily.

Coal is plentifully available in the two leading greenhouse gas emitters, the US and China – both of which are sorely feeling recent price volatility and uncertainty in the availability of oil. So neither country can sensibly build more oil-fuelled plants, and both are far away from sources of natural gas. China produces and uses almost half the world's coal, the US almost a fifth: nearly two-thirds of the world's total between them.

But those of us who would like to see much greater use of green energy such as solar and natural gas have to realise that the technology is not ready, costs are too high and needed infrastructure is not in place. Periods of low wind affect all wind turbines in an area and cloudy periods all solar facilities – and yes, both can occur together. So it's hard to know just how much (inefficient, expensive) storage you need with these sources. Large, extensive, "smart" electricity grids are needed to get renewable power to the right place and ameliorate availability problems; but they're not yet in place, nor even firmly committed to.

All of these problems can be solved, but only with a lot of expense and a fair amount of time. Whereas with a fossil fuel-powered power plant, you can turn it on and off as needed, and fuel supplies are more reliable. This is hugely beneficial to stable supply of power to home and business users – that is, to all of us. So, with oil supplies unsteady and prices unstable, and gas off the table in many countries that lack the infrastructure to bring it in efficiently, coal can seem the only solution.

There are high hopes for "clean coal" burning through carbon capture and storage (CCS) technology, but those hopes are gradually fading as research continues. CCS seems to be a very difficult, expensive and, ironically, energy-consuming solution, needing many more years of research and development before it can potentially be put into use.

On balance, and with great sympathy for power utilities worldwide and their customers, any serious approach to saving the planet requires that new coal plants be banned within just a few years. (Not subjected to carbon markets, offsets, taxes or trade-offs – just banned.) Failing a miracle on the CCS front in that timeframe, a programme to retire them must be energetically pursued.

The rest of the world needs to save the Chinese, in particular, from themselves by then finding ways to make coal use unprofitable. One mechanism is the International Standards Organisation (ISO) process. There is already a set of ISO environmental standards called ISO 14000. ISO 14000 can be used to specify the makeup of green supply chains, including the use of relatively clean power – perhaps, at first, specifically excluding coal power only. Incentives could also be created for retiring coal plants.

Al Gore and, surprisingly, Ed Miliband – at the time, UK Environment Secretary – are the most prominent among those who have called for the use of civil disobedience against the construction of new coal power plants. I sympathise with those finishing up construction of plants approved years ago, and which generate much-needed power. I would like to see 2013, perhaps, declared as the first unlucky year for new use of coal, and any plants built on or after that date – anywhere in the world – would be the site of protests, downstream boycotts and other action for years to come.

This is less dramatic and, perhaps, less exciting than trying to stop new coal plants from being licensed today. But climate change is a huge problem that deserves a thoughtful response that will "stick", and that can gather widespread support, not just a knee-jerk reaction. To me, pausing to gain support, and to allow alternatives to mature a bit more, before taking action against new coal plants is an example of going slow (briefly) to go fast (from then on).

Close behind winding down the use of coal in urgency is stopping deforestation and beginning reforestation.

Forests are the lungs of our planet and we are quickly losing lung capacity. Growing populations needing land to live on and grain, the need for even more land to support animals for increased meat production, the ongoing purchase of tens of millions of new cars and their need for roads and parking, and of course the need for timber, all press on forest land.

At the same time, warming and changes in rainfall patterns are making formerly productive forests marginal and destroying marginal ones. Deforestation by people and damage to, and destruction of forests by climate change are pincers that are crushing forests between them.

We desperately need to preserve the remaining healthy forests we have. This is necessary for any fight against climate change, let alone against runaway climate change. Reforestation is our only proven tool for taking greenhouse gases out of the air. We need to preserve forest today, to slow climate change, and to provide a base for reforestation to fight runaway climate change tomorrow.

Deforestation, like much coal burning, is taking place in just the regions that are seeing rapid population increases and rapid economic growth. To take two examples: in India, economic growth struggles to keep pace with population increases, while in China, with its draconian one child per family policies, growth is gradually enriching the population. Each set of circumstances has specific disadvantages from a deforestation point of view – desperate, starving people cut forests down quickly, but so do ranchers seeking grazing for cattle and engineers building highways.

Current policies in developed countries encourage deforestation and other environmental problems in the developing world. Tight environmental rules combine with higher wages and other regulations to raise costs in the developed world, allowing less-developed countries to underbid by enough to cover shipping costs, supply chain problems and delays – but only by despoiling their environments and ignoring the health and welfare of their workers.

The most practical mechanism for preventing deforestation might be a commitment to stopping it in developed countries and direct payments for preserving forests in developing ones. This might go to the extent of buying or leasing the land the forest occupies.

The current trend is toward an overall, market-based approach to climate change prevention based on carbon trading and selling offsets. This allows one kind of damage to be traded off against another, but it may not get us where we need to go fast enough. Perhaps specific protections for forests are needed, including the use of international standards and consumer pressure on those responsible for the forests.

The third short-term opportunity is to begin the process of reforestation. Reforestation is the flip side of stopping deforestation, a critical step toward moderating climate change and beginning to address runaway climate change.

Goal 4. Zero tolerance for CO_2: Power

The first step in having zero tolerance for carbon emissions is having zero tolerance for carbon emissions from power generation. Carbon-free power is power that doesn't produce greenhouse gas emissions. That rules out, in order of undesirability, peat, shale oil, coal, oil and natural gas.

Even the use of draft animals and human effort are powered by food that, in being grown and transported, consumes power that is generated, in today's world, mostly by fossil fuels. (A recent analysis showed it can generate fewer greenhouse gas emissions to drive a mile to the store in a small car than to walk, because the production and distribution of the food that fuels walking is itself so greenhouse gas-intensive.)

The main sources of carbon-free power that are in use today, with varying prospects of improvement for the future, are wind power, hydro-electric power, geothermal power, nuclear power, solar power and tidal power.

The first priority is to understand just how important carbon-free power is. People's activities cause greenhouse gases to be emitted by a chain of events:

- Start with a person
- That person makes money
- That person spends money on goods and services
- Supplying goods and services causes energy to be consumed
- The energy used causes emissions.

So there are five ways to reduce emissions:

- Reduce the number of people
- Reduce the amount of money each person makes
- Reduce the amount of money each person spends
- Reduce the amount of energy used by the goods and services each person spends money on
- Reduce the emissions generated by the energy used.

Every step in the above list of possible reductions gets advocated by some along the way, and as the urgency of reducing emissions increases, various measures get pushed with increasing urgency: "Reduce the surplus population", as Dickens' Scrooge put it; live more simply; consume less; conserve energy. All are urged on people – and all may be important for the overall task of saving the planet. But environmental advocates can come across as carping, negative, weird or even inhuman as they push these steps, out of context, on various decision makers and on mass audiences.

But the simplest, best and, given the scale of reductions needed, only effective way to reduce emissions is the last one: to reduce emissions generated by the energy used. That is, carbon-free power.

With carbon-free power you can save your people, let them get rich, consume goods and services and even, once polluting power is out of the picture, waste energy; from a global warming point of view, you're golden. (Though we have many decades in which energy efficiency will be crucial before we get there.)

Carbon-free power can even help address agricultural and water problems. By powering drip irrigation, carbon-free power can help reduce water usage, and can fix nitrogen for fertiliser without adding to greenhouse gas emissions.

And carbon-free power will help us avert what might otherwise be a disaster. Peak Oil may already be upon us, threatening ruinously high prices for this vital resource. It may be followed in a few decades by Peak Coal. Even if it doesn't "peak", coal use may become increasingly restricted as greenhouse gas emissions begin to be controlled. From whatever cause, diminishing oil availability and diminishing availability or usability of coal will be an economic disaster, unless carbon-free power is ready to take up the slack.

Carbon-free power is the most positive and liberating step we can take for humanity, freeing us from fossil fuels before we run out of them, along with the original impetus – to avoid the worst consequences of climate change.

So what's the best way to get carbon-free power? The first step is a flexible and "smart" electricity grid, the main element of which is relatively new technology that allows power to be moved long distances with little loss. A "smart" grid can also support buying power from a home with solar panels, for instance, one minute and selling power back to the same home the next. Smart grids support charging more when power needs are greatest and less when demand is lower.

A truly effective smart grid is large, even continental – say, North America- or Europe-wide – allowing power sources and power users to be located where it makes the most sense. In North America, for instance, dams in Canada and solar plants in Arizona could both supply air conditioners on a hot day in Chicago.

A large smart grid also allows amelioration of the worse problem with most renewable energy sources, such as wind and solar, which is that they often aren't running at peak output when you need them. Grids that cover several time and climate zones are more likely to be able to mix sources and trade off production with need.

Much-improved power storage and fossil fuel plants kept as a ready reserve are among possible solutions that will need to be developed further. (This last sounds bad, but those of us pushing for greenhouse gas reductions need to keep reminding ourselves that brownouts and blackouts must be absolutely minimised with new technology; energy always means productivity and comfort as well as, often, life itself.)

A smart grid allows for maximum flexibility in power sources and in reducing power use. But where should the power come from? Solar power seems the best starting point. It's generated by nuclear fusion, the greatest source of energy in the universe, taking place conveniently far away, within the sun.

There's 10,000 times more energy falling on the Earth's surface every minute, as sunlight, than we use. So with even a mere, say, 10%

efficiency in gathering sunlight, and just 10% efficiency in transporting, storing (where needed) and putting it to use, we would need to cover 1 percent of the Earth's surface with solar panels. Luckily, much higher efficiencies are already being achieved, steadily reducing the sheer acreage needed.

Every percentage point improvement in efficiencies – from the capture process, from better batteries and other storage, from better transmission grids – helps "mine" this resource more effectively. Recent estimates are that a solar grid 100 miles on a side could supply electricity for all of the US.

Solar power is also most plentifully available where it's needed most – in equatorial and near-equatorial countries and regions in line to suffer most from climate change, and where most of the poorest people live. With solar power, these unfortunates of climate change can get plentiful power cheaply and, with extensive smart grids, sell the surplus on to developed countries. Greater use of solar power gives a potential lifeline to those hardest hit by climate change.

Other carbon-free sources, except nuclear, are ultimately derived from solar power – as are fossil fuels, of course. Tidal and wind power, for instance, therefore have less overall energy potential than solar. However, they should be considered in higher latitudes where solar is weak and opportunities for specific alternate technologies are strong.

Countries that have at least one strength in technology and deployment will be able to bargain effectively for the pieces they don't have. For instance, the US should arguably make large investments in solar power; the UK, in tidal power, both using smart grids to get the power produced in remote areas to where it's needed. As two of the world's largest greenhouse gas emitters over the centuries and, on a total and per capita basis, today, both countries also need these "chips" for deals to pay their share in helping repair damage and reduce emissions in the developing world.

For just one example, various forms of virtualisation – using teleconferencing to replace many business trips, teleworking, Internet shopping, getting more of one's entertainment at home over the Internet – all promise to greatly reduce greenhouse gas emissions from travel. But the data centres that power all this are themselves becoming big emitters through their power use, threatening, in aggregate, to rival air travel and air freight as polluters in coming years. Drastic reductions in emissions for electrical power will eliminate the problem.

The development of carbon-free power sources and the deployment of large and growing smart grids is what chemists call the "rate-limiting step" in any and all efforts to reduce greenhouse gas emissions. With them, a nearly complete transformation is possible; without them, steps such as greater conservation are only nibbling at the edges of the problem.

Al Gore has proposed a ten-year plan for the US to move to carbon-free production and distribution of electricity. His plan is exactly the kind of effort that will give not only the US, but the world as a whole, a fighting chance in the battle to reduce emissions. Germany and the UK are two other countries that have great opportunities to lead in such development today and develop valuable assets for the future.

Goal 5. Zero tolerance for CO_2: Transport

Transport is a huge and growing issue in emissions. Renewable energy sources and smart grids have the potential to radically reduce the carbon intensity of much of daily life. But moving to near-zero emissions from transport will take additional steps, many quite radical.

The lowest-hanging fruit for developed nations is the use of cars and trucks for transport. With carbon-free electricity, electric vehicles become a marvellous solution. The energy security advantages of steadily diminishing oil use will only encourage countries to, well, encourage such transitions.

Mass transit is in some cases already electrified – the London Underground and the streetcars of San Francisco are two famous examples – or fairly easily moved to hybrid power, with the electric part used more and more.

In developing countries, the challenges are harder. Less likely to be leaders in the development of new technologies for carbon-free power generation, they must wait for such technologies to emerge, then either pay or negotiate for access to and deployment of them.

Developing countries also face severe land use and water use constraints as they choose a way forward. It's possible for developing countries to follow a different development path from the very-high-emissions, high car ownership and high land use American model, or even the merely high-emissions, moderate car ownership and more moderate land use European model. But so far both China and India, the two bellwethers, seem intent on pursuing an American approach that's physically impossible to fully implement, given their population densities and their need for arable land.

Developing countries will need to find sustainable transport policies to support their already dense and fast-growing populations. They'll need help from developed countries to implement carbon-free power and use it to support their transport networks.

While most of the problems in decarbonising land transport are solvable, a notable exception is the sheer consumption of land for roads and parking, which causes deforestation and reduces the land available for reforestation and agriculture. The competing needs for land will require that road space and car ownership be limited and, in some countries, reduced.

George Monbiot points out in *Heat* that a single bus can take many cars off the road. Shared-use, low-emissions cars in cities greatly reduce emissions and parking needs.

But all this potential for progress in land transport doesn't offer much help for two large and fast-growing contributors to greenhouse gas emissions: sea and air transport, each responsible today for about 4% of greenhouse gas emissions, but both set to increase steadily.

These forms of transport are the underpinnings of otherwise desirable increases in trade and travel. Trade – which drives nearly all shipping (both sea and air freight) and a high share of air travel – has been growing at twice the rate of economic growth, with only a brief slowdown for the current recession. This implies steep further increases in shipping and business travel, with personal travel following along as the world gets ever smaller. The first-order approximation is that both shipping and air travel – and their greenhouse gas emissions – will tend to increase at twice the rate of economic growth as well.

As electrical transmission and land transport are de-carbonised, with more and more of the world's energy usage driven onto increasingly carbon-free "smart" electrical grids, the rapid absolute increase in shipping- and travel-related emissions will become an even greater increase in their share of total emissions. Put another way, sea and air transport emissions threaten to undermine hard-won progress elsewhere.

Shipping is increasingly recognised as a major contributor to greenhouse gas emissions. It's only in the last few years that the total weight and distance has really been added up, and some representative "tailpipe" measurements made of emissions. The results are startling; because ships run on the cheapest, heaviest sub-diesel fuel, and because their emissions are unregulated, they pollute heavily, out of proportion even to the massive volumes of freight they transport and the long distances they travel.

De-carbonising shipping is really tough. International pollution controls can gradually shift ships to less polluting fuel mixes. Nuclear-powered ships, already in military use for decades, could help, but it would take many decades to scale this up – even if proliferation and safety concerns for many tens of thousands of ships, most relatively small, and nearly all operated in the least expensive way humanly possible, and registered in the most forgiving jurisdictions available, could be addressed. (The potential for converting oil super tankers to cargo ships, as oil use on land drops, and running these huge ships more cleanly or even on nuclear power, could help solve part of the problem.)

Even with these changes, shipping will need to be shifted over to land transport where possible – containerised shipping helps here – and made quite expensive where it isn't. If growth in sea transport can be stopped, in favour of more use of land transport, while the ship stock is partly converted to lower emissions and perhaps partly nuclearised, the problem will at least be gradually reduced, though not revolutionised as can be done with land transport.

Air transport is even worse. Whereas sea transport can, with wrenching changes, be significantly improved, significantly converted to land transport, and perhaps even sped up in the process, air transport (of freight, packages and passengers) offers speed and, for long trips, convenience advantages that land transport can't match.

In *Heat*, George Monbiot cited air travel as the one area where he couldn't manage his carbon footprint gently downward; he would simply have to avoid air travel at almost any cost. The same may well hold true for society as a whole. The most wrenching change we may all need to make is to dramatically reduce use of air travel. Business class and first class, with all the extra legroom and chairs that turn into beds, have two to four or more times the impact of a seat in steerage – I mean, economy class.

It's technically possible to nuclearise air transport, but I can't seriously propose this as a solution given today's near non-use of this technology, its potential dangers and public attitudes toward nuclear energy. And aircraft, even more than ships, can't run off a de-carbonised power grid; they need high-density portable fuels. In other words, fossil fuels, at least today – perhaps much-improved biofuels tomorrow.

Added to this is the unique nature of air transport in that planes fly high in the atmosphere, causing damage of a type that may not yet be fully understood, but is certainly out of proportion even to its high greenhouse gas emissions.

So cutting air transport may be the most urgent and wrenching change we make in fighting climate change. A programme to accomplish this would require vastly improved videoconferencing capability and quick implementation of all the improvements above in land and sea transport.

Air travel could be redeveloped to mostly handle over-sea transit, with a few key points – such as, for example, London, Paris and Madrid for Europe; Miami, Boston, New York City and Washington, DC for the US East Coast – interfacing between a vastly reduced air fleet and vastly improved land transit. Fast, convenient point to point air freight and air travel as we know it today would become very expensive and infrequent.

Air transit is a great leveller and door opener for new business opportunities and personal growth. In *Heat*, Monbiot additionally cites the invaluable role of "love miles" in connecting far-flung networks of family and friends. All of this is real and valuable, and severely reducing it will be a huge – but, on current evidence, necessary – sacrifice in the battle to reduce greenhouse gas emissions.

Goal 6. Back to 280ppm and pH 8.2

Those who have asserted that climate change might be runaway have been criticised as "defeatist". But that's not right at all.

It seems there's a split today between those who want to reduce emissions now and those who want to use technical means to remove greenhouse gases from the atmosphere directly. Each side talks as if their own approach is practical and affordable, and the other approach is dangerous, ruinously expensive and stupid.

Actually, both approaches – reducing emissions quickly and removing greenhouse gases directly from the environment – are quite impractical and very expensive. Which is why we have to use both at once.

Fairness and safety require that we go "back to the future". Atmospheric greenhouse gas levels must be cut to pre-warming levels of 280ppm. Dissolved carbon in the seas and fresh water must be removed to return acidity to pre-emissions levels – a pH level of 8.2, rather than the pH level of 8.1 that's been reached today. Only active steps to remove greenhouse gases from the atmosphere and waters can accomplish this, and only reductions in emissions to near zero can put us in range of doing so.

We have to cut emissions because we can't afford to keep cleaning them up – and we have to clean up what's already there because we've left it to too late to begin cutting them. There is probably a small margin where we might cut emissions a bit more slowly if removal technology comes on quickly, or cut more quickly if not, but we will still have to do both very aggressively.

Accepting that carbon sinks are tipping obligates us to do more, not less, and to do it faster, not slower. To give up at this point would be a truly historic abdication of responsibility.

The only safe and sure way to remove greenhouse gases from the atmosphere is to stop deforestation, allowing current greenery to continue as a carbon sink, and to begin reforestation, moving carbon out of the atmosphere and into plant life.

This effort will have as a crucial side effect the preservation of many species of plants and animals, securing the benefits of at least some of the biodiversity currently found on Earth to ourselves and our posterity.

But it's important to spell out that reforestation is only an answer if emissions are cut rapidly toward zero. The Amazon rain forest contains the equivalent of 15 years' emissions. There might be room to add the absorption equivalent of one or two more Amazons on Earth. This capacity can't be wasted on partly countering some small share of ever-growing, "business as usual" emissions; emissions must be cut sharply, and reforestation used to reduce overall greenhouse gas levels.

The most effective places to stop deforestation and begin reforestation are in tropical rainforest areas, as these support the densest growth.

Reforestation is hugely difficult. Land is under tremendous pressure for use to support more growth of grain, more biofuels and biopharma, more stockyards and more cars. This will only increase as the population grows and becomes increasingly industrialised, and as large swathes of today's productive land are impaired by greater heat, less reliable rainfall and snowmelt and the poisoning effects of pollution.

Today's small environmental preserves can scarcely be maintained in the face of human pressures in wildlife and on the land; establishing and maintaining huge new preserves will be wrenchingly difficult.

There are additional, complementary means under investigation for removing CO_2 from air and water; for instance, certain minerals, unearthed and treated appropriately, absorb quantities of CO_2. Such means face a variety of technical problems, but there is also a potential conflict of interests that may manifest itself as a legal issue.

Basically, the ground belongs to individual nations, but the air belongs to everyone. A nation's right to manage its forests, for instance, is difficult to dispute. And one country can pay another to reforest.

But deliberate, artificial steps to remove CO_2 from everyone's air run into some of the same legal objections that would arise if a country took deliberate steps to pollute everyone's air – at an extreme, such could be considered an act of war.

Not all nations will have the same opinions, interests or goals in slowing, stopping or reversing climate change, with Russia and Canada among those who receive a net benefit from warming.

It may be both a necessity, and extremely difficult, to get broad and explicit agreement on technical measures to reduce greenhouse gas levels in the atmosphere. There may be room for argument that a given nation has the right – perhaps even the obligation – to remove the pollutants which that nation put into the environment. But that's an argument, not an agreed principle.

This worry may hang over most efforts to reduce greenhouse gas levels and reverse the effects of climate change. But it must be resolved or evaded, as steps to remove CO_2 from the atmosphere are badly needed.

Getting CO_2 out of the oceans is also important as part of an effort to de-acidify them, so highly productive coral reefs can survive and so marine food chains can be preserved. It's hard to imagine anyone arguing against ocean clean-up.

There will also be the need for technical fixes to reduce the impact of sudden "blooms" of methane from currently frozen peat and, potentially, from clathrates. Unstopped, the sudden release of methane could become self-sustaining in a specific region at the same time that it affected temperatures globally. Methane blooms may need to be fought in a similar manner to forest fires, and the means must be developed to do so.

Goal 7. Lower the temperature

Reducing greenhouse gas emissions will be slow and expensive, though we at least know roughly how to do it.

Removing greenhouse gases from the environment is also slow and expensive, and we don't know yet how to do it. In addition, getting the needed agreement and assembling the required political will is a truly daunting prospect.

But the final step is also slow and expensive, and unlikely to find the needed political will and agreement at all; it will work against too many people's interests, from shipping companies transiting newly ice-free Arctic routes to senior citizens in New England newly freed from high winter heating bills. But we need to cool the Earth back down below pre-warming temperatures for a while as part of a return to the status quo ante.

It's surprising that there's no discussion of cooling the Earth in most mainstream discussions of battling climate change. There's a discussion of reducing emissions vs. removing carbon by technical means – as if these were mutually exclusive – but little discussion of cooling the planet, despite recognition that climate change might go runaway.

As it is, climate change is already runaway, and we will need to cool the planet to halt it and reverse it. There's too much CO_2 and methane embodied in trees and entombed beneath permafrost – on land and under the seas – that's at risk of escaping into the atmosphere, and too much otherwise permanent damage to reflective snow and ice surfaces, and to glaciers, for us to do anything else.

So cool the planet we must. A programme to pursue this must be begun immediately; it may end up being referred to as the "Big Chill", if you'll excuse the reference to the feel-good 1980s movie.

In discussing cooling the planet, in addition to a catchy name, one has to have a target. But why do I recommend that we set a return to "pre-warming", to use a perhaps inevitable shorthand name for it, as the benchmark?

We have to set some kind of target temperature. While any given target will have winners and losers, the pattern of human population on Earth is aligned to pre-warming temperatures and sea levels, and the crop-growing patterns, water flows and so on that have existed for many centuries.

Anything hotter than pre-warming temperature levels puts the lives of hundreds of millions at risk through sea level rises, impacts of fiercer weather and the continuing risk of runaway climate change. Even in cooler climes, the same hard freeze that raises heating bills and forces road and school closings also kills off pests that otherwise are left free to spread disease, among humans as well as among trees and wildlife.

So, while a fuller discussion is needed, the result is likely to be that humanity as a whole is best served by a return to the pre-warming environment. And to get there, we need to initially overshoot – to cool the planet even below pre-warming temperatures.

Why? To restore glaciers worldwide, especially the North Pole's ice cover, and to ensure that other carbon sinks, already under pressure, don't tip. Just as it took several decades of above-average temperatures to tip the North Pole's ice cover toward destruction, it will take several decades – or perhaps several centuries – of below-average temperatures to restore it.

Only when we're certain that the North Pole's ice cover is fully on track toward restoration, with greenhouse gas levels and other carbon sinks in balance, will we be able to let the pressure off.

Note that I'm not advocating that we, out of nowhere, start manipulating the temperature; we already are. I'm advocating that we manipulate it purposefully and with a goal in mind rather than stumbling about doing it by accident, with the kind of unforeseen consequences we are experiencing today.

So how can cooling be accomplished and adjusted? There are actually many possibilities that will need to be compared and costed. The initial decisions may not necessarily be the final ones, as we learn more and as costs and benefits become clearer.

The first principle will have to be borrowed from the Hippocratic Oath for doctors: first, do no harm. Any means that has a risk of devolving into excessive or even runaway cooling has to be avoided.

With that in mind, we can take a look at possible solutions, starting with the most conservative first:

- **Direct replacement per carbon sink**. For instance, replacing the ice floes that make up the North Pole's ice with something else white that floats; white floats made from biodegradable materials would be one solution. This would be a direct and scalable replacement for the missing sea ice. Surface cover on land could accomplish the same purpose.
- **"Safe smog"**. It's believed that largely Western smog helped slow warming by reflecting sunlight in the 1960s and 1970s and that largely Asian smog is doing so today. A safer form of smog could be invented to give us reflective cooling without respiratory problems.
- **Space shields**. Reflective chaff or various kinds of satellite-deployed reflective shields could replace the reflective effect of North Pole ice until it returned.

The technical issues in cooling the Earth are likely to be huge, as are political issues. There will be strong opposition to risks, and advocates – perhaps very forceful ones – of retaining any of several different degrees (no pun intended) of warming. Acting on this without 100% agreement would be hard to justify in law or in fairness, yet it will have to be done. All of which could lead to military confrontation, even nuclear sabre-rattling or worse.

But we've reached the point where not actively cooling the planet is not an option. Runaway climate change puts almost every species we know of at risk, as well as much, most, or even all of humanity. We have to stop it, and that means reversing it, at least in part and preferably – and most safely – completely.

Yet this is describing a balance of probabilities; there is no risk-free solution. We've rolled the dice with our greenhouse gas emissions; from here, even the best outcome we can manage will include great expense, great difficulty and great dangers.

Mechanisms for encouraging compliance

I've rather blithely laid out a programme of action that would change our world – albeit while preventing the far greater changes that runaway climate change is on track to bring. How on Earth can any of this be accomplished, let alone all of it?

I believe there are two initial steps that will, if taken, make a huge difference. The first is for a critical mass of educated and public opinion to realise and accept that we are indeed well into runaway climate change.

This is a paradigm shift of immense proportions, and those who have won laurels for their work on climate change so far will not all admit to error quickly. They will be backed by those who have opposed recognition of, or action on, even the relatively easy-to-swallow 2007 Consensus.

This latter group, especially the segment that is or has been in positions of power, realises they are beginning to risk looking to history not unlike war criminals. Yet many of them will go to their graves arguing vociferously that they were right – even if an awful lot of people suffer and die as a result of continued inaction.

But the facts of runaway climate change are clear and obvious all around us. So this may end up being – it very much needs to be – one of the faster and easier paradigm shifts in history.

After enlisting thought leaders, next comes getting the IPCC on board. The IPCC process is so influential, and so many of the needed scientists are bought into it, that by far the quickest and most effective way forward is for the next IPCC Report, due to be published in 2014, to offer an accurate reflection of the situation we face and the risks going forward. If it doesn't, a long and messy fight will have to follow to try to correct it, and valuable time and focus will be lost.

With leading-edge opinion on board, and an accurate and insightful 2014 IPCC Report (or alternative scientific resource) on the books, the entire climate of opinion should shift.

To lay the groundwork for this to happen, and to take advantage when it does, will require efforts at every level of action – by individuals, organisations and governments.

The trickiest part of climate change solutions is the need for nations to cooperate. But one language that nearly everyone responds to, in all countries and at all levels, is money. An increase in demand for green supply chains, as mentioned above, may be critical in motivating action.

This would give individuals, businesses and even governments a way to "vote with their wallets" for making growing sectors of the economy carbon neutral. Eventually, non-green products could even be penalised or banned in trade agreements – "free" trade having proven to bear quite a few costs.

This would gradually remove cost advantages that some companies get today by locating in places with lax environmental regulation – advantages that will otherwise only grow as some countries take climate change less seriously than others. With the regulatory playing field levelled, and as wages gradually rise in today's low-wage countries, the main competitive advantages for companies will become reputation (called "branding" today), efficiency and proximity to resources and customers – all advantages that are environmentally positive.

The main things an individual can do today to help address runaway climate change are:

- Support current efforts to counter climate change. Almost everything being done under the current paradigm will be useful under the new one as well. The Gore book continues to be a good primer here.
- Try to get people you know to understand that climate change is already runaway. This book is one resource, but there will no doubt be others.
- Look for ways to influence the IPCC process and to support politicians interested in acting on climate change. The IPCC is both scientific and political; exercise any influence you can muster on both.

It's simultaneously very early in a new paradigm and very late to take effective action, making it hard to know what best to do. Educate yourself now so you can be ready to act effectively as opportunities arise.

Chapter 10. How Nations Can Respond

> In this chapter
> - A template for national responses
> - Recommendations for the UK
> - Recommendations for the US

Each nation has different challenges – with some regions verging on crisis already, and others likely to follow – and opportunities in facing climate change as currently understood, a host of more or less related environmental and demographic problems, and the swing from current complacency to recognising and battling runaway climate change.

In most cases, it's at the national level that the decisions will be made which will see the world as a whole succeed or fail in containing and reversing runaway climate change.

The first obligation of a nation is defence, and the number of threats nations are tasked with defending against – not just military but economic, social and environmental – continues to rise, putting nations under increasing strain.

It will take widespread recognition of the dangers posed by runaway climate change and related problems to inspire nations to take the appropriate steps to deal with them.

Once recognised, nations will have to come up with a plan for dealing with them. The key values I recommended in Chapter 9 for battling runaway climate change and related problems include honesty, fairness and making saving human lives the clear #1 priority. These are not just feel-good aspirations but vital tools for garnering and maintaining public support.

The key overall goals I recommend are stopping deforestation immediately, cutting global emissions to zero by 2070 and pursuing geo-engineering solutions to reverse runaway climate change.

For individual nations, the steps I suggest in line with these values and goals are:
- Plan how to address runaway climate change as a global and as a national issue
- Rally popular support

- Control national borders
- Plan for rising seas
- Plan for changes in weather
- Seek self-sufficiency in water and food
- Seek self-sufficiency in energy
- Develop a leadership position
- Provide for the common defence
- Strengthen the safety net.

All of these steps are needed, and each reinforces the other. A country has to have a plan for addressing runaway climate change as a framework for other actions; it can't enact its plans without popular support; and it can't provide for defence if it's too exposed in relation to supplies of water, food and/or energy, inter-related as they are.

A template for nations

The first thing that any country has to do, with runaway climate change on the horizon, is to understand its situation – how it will be affected directly by rising seas, by the most likely changes in longer-term climate, annual weather patterns and extreme weather, and how neighbouring countries are likely to be affected as well. Each country also has to determine for itself how and at what point runaway climate change can be stopped and reversed. All this can lead to preventative steps today that save a great deal of trouble later. It will also inform the country's stance in (runaway) climate change negotiations, energy policy and more.

Countries should undertake an initial, categorical study and establish a plan for updates, perhaps keyed to the IPCC reporting cycle of roughly six years – but not bound by IPCC reports. The computer power needed to do detailed regional and national projections is expensive; most countries are likely to want a national capability for simulation studies, which can be powered by anything from supercomputers – a small one costs only $4000 – to citizens' PCs crunching numbers in their spare cycles.

The study period will have to include assessing the likelihood of runaway climate change, its impact on the country as it unfolds, and how that country might influence and participate in efforts to stop runaway climate change.

It's only as this detailed work is done, one nation at a time, that runaway climate change will become real to planners and politicians. When the findings are made available, and energetically communicated to the public, then people can become better informed and more engaged.

At this point a nation will need to set national goals for areas it can control, such as energy self-sufficiency, and how it will lead or participate in reversing runaway climate change. The more broadly based and inclusive the goal-setting exercise, the more chance the goals will have of being met.

It's only when people understand what runaway climate change means for them, their country, their families and communities – now and in their children's and grandchildren's lifetimes – that a really effective global discussion can begin. Only then can they be ready to make bigger sacrifices and invest time, energy and money to prevent it, as individuals, as members of communities and as citizens, both of their respective countries and, in a broader sense, of the world.

Rallying popular support is a difficult but vital part of responding to climate change, even in autocracies. (China was seeing thousands of demonstrations a year even before the credit crunch hit; autocracy isn't necessarily what it used to be.) Information and planning are the first steps in ensuring popular support.

People need to be kept informed and involved in decisions. If a shocked and dismayed population turns to rioting or open revolt, governments can collapse just when they're most needed.

People can endure a great deal when things make sense to them. But climate change is very subtle and seemingly random in many of its effects. Adding sacrifices to confront or reverse runaway climate change to the deprivation and difficulties that will occur as it unfolds is asking a lot.

Winston Churchill did a masterful job of preparing a population for war at a time when the enemy was yet to strike, and his people then endured stoically and heroically under attack. Facing the same enemy, the Soviets, ill-prepared, suffered huge losses – and saw the loyalty of large swathes of the population in doubt – before finding their footing.

Trying to match Churchill's experience, credibility, focus and stirring rhetoric is more than anyone should be expected to manage, but a calm, thorough educational effort now will pay great dividends later.

The current credit crunch and recession are actually the best time to lay the groundwork for an uncertain future, giving meaning to the sacrifices people are already making and the further ones they're likely to endure – and increasing the odds that the stoic model set by the civilian population of World War II Britain can become the norm in an increasingly challenged world.

"Regaining control of our borders" is a favourite conservative, and sometimes nationalist or racist hobbyhorse, but it will be a necessity for even the most progressive government as climate change kicks in.

Controlling one's borders doesn't mean stopping immigration or preventing greater diversity in a country's make-up. It means managing immigrant flows to a democratically agreed level and preventing basic services from being overwhelmed.

It's usually the poor who suffer first and most from uncontrolled immigration, finding themselves thrown into competition for lower-paying jobs, for housing, school places, medical care and social services. We have already seen that climate change poses great challenges for poor people, such as flooding and sudden upsurges in food and energy prices.

It's unfair to expect the poor, in particular, of one's country to deal with large and sudden – often even uncounted – surges of immigrants. The European Union will have a particular problem here, as it has removed all immigration controls among member countries. The EU's southern tier – Greece, Italy, Spain and Portugal – is due to be annexed by the expanding Sahara Desert as it "jumps the Mediterranean." (Already, Spain is in drought, and Greece has suffered record-setting forest fires in recent years.) Less-affected EU countries may experience unprecedented surges of immigrants. With the large migrations and the strains expected even on the best managed and most fortunate national economies, life itself may come to be at stake for both established residents and incoming migrants.

What about the rights of migrants fleeing climate change? In a world organized by nations, there isn't a human right to move into someone else's country (except among EU countries), simply a struggle between those who have a tenable home base and those who want to move into it. People need to be helped where they are as much as possible. Climate change will strike so severely and unpredictably that it will otherwise be all too easy for migrants to be chased from one marginal toehold to another, suffering escalating losses along the way.

Resistance from established residents – even those who are earlier migrants themselves – could become fierce. We can't afford to see refugees by the millions subjected to scenarios resembling Napoleon's retreat from Moscow. A seemingly open, or poorly patrolled, border could simply tempt people into moves that will be counterproductive for everyone in the long run.

Even with the best efforts, climate change and related demographic and environmental problems are likely to create tens of millions of refugees, and it's likely that the more stable nations will end up agreeing to accept large chunks of refugees. Nations with control of their borders can do a better job of absorbing refugees with less disruption than those with leaky or more or less open borders. A moderate, managed approach – only possible with controlled borders – is far more sustainable than uncontrolled surges.

Countries also need to plan for rising sea levels, perhaps a full metre by 2050, perhaps another metre by the end of the century. In some areas, a metre of sea level rise can mean anywhere from many metres to a mile or more of ocean encroachment. There is not yet an established mechanism for an orderly retreat from the sea or for making decisions as to what areas to defend, and for how long, but one needs to be created; without such a framework, the decisions are just a grinding, exhausting, expensive series of legal, political and financial power struggles.

In coming decades, we may have to expect a new Katrina-level disaster in a major world city every few years. Katrina generated hundreds of thousands of refugees from New Orleans, some of whom were still living in trailers years after the event, and many of whom will never return home. Imagine the effects on the world of repeated disasters, of Katrina scale or greater, usually happening to far poorer countries than America.

The need to plan for rising seas will accompany a need to plan for, and cope with, severe changes in weather. Extreme weather events, often worsened by ever-higher sea levels, will get the most attention, but more gradual shifts in rainfall patterns will matter just as much over time. Yet distinguishing among and combating short-, medium- and long-term weather changes, with extreme weather ever more common and the progress of runaway climate change uncertain, will vary between difficult and impossible.

All this change will make providing adequate harvests extremely difficult, at a time when countries need to work toward self-sufficiency in food and water. (Food self-sufficiency is firstly a function of water adequacy, plus using enough productive land to get needed yields in spite of adverse weather conditions.)

Countries (and states, provinces, regions and cities, for that matter) that depend on river flows stand to lose out to those upstream, who will have first crack at threatened supplies. Some populations in currently vulnerable countries and regions may simply be unsustainable in their current locations; southeast Australia, the American Southwest and Egypt are among many potential victims.

So countries need to ensure water self-sufficiency, firstly on a national basis but by region and locality within their country as well. This may require expensive long-term water purchase agreements, massive irrigation works, new, energy-intensive desalination plants, even preventative shifting of populations – which will seem an extreme measure until compared to the consequences when serious water shortages hit.

The kingdom of Tuvalu has begun a 30-year effort to resettle its population of several thousand people to New Zealand, moving 75 people a year. On a larger scale, China has recently discussed plans to relocate more than 10% of its population in response to desertification – more than 100 million people. These will be far from the last such efforts.

Countries need to be able to grow enough food to support their populations, at least in terms of total calories if not variety. (If quantity is sufficient, variety can be traded for among nations that have adequate food.) World grain markets are likely to see shockingly high prices as shortages develop for grain as food, as animal feed and as fuel stocks.

Shortages will hurt many directly and the resulting high prices will quickly hurt everyone worldwide, especially the poor. The only defence for any one country will be adequate supplies of food for its own citizens, enabling governments to supply at least basic rations of grain, fruit and vegetables – less so meat, which may become more of a luxury – for a reasonable price.

Energy self-sufficiency – and in "green" energy – is another necessity. Every country will have different opportunities among solar, tidal, wind, and nuclear power sources, all of which will shift in cost and practicality as development accelerates. The best policy may be to guesstimate the future mix based on fundamental factors like latitude, miles of coastline and wind exposure, provide incentives to get development started toward that mix, then be ready to shift tactics as opportunities open up and problems emerge.

It may be possible for some countries to seek energy assurance rather than energy self-sufficiency through long-term deals, construction of infrastructure such as pipelines and multi-country electrical grids, and so on. The same may be true for other vital needs such as food.

But, while we enjoy increasing benefits from freer trade and other multi-national arrangements, we are heading into this crisis without effective means of enforcement for international agreements – short of military force – and the fundamental unit of political control remains the nation-state. Prices are unpredictable, especially for fossil fuels, as are penalties that might be agreed – or even imposed, on weaker countries – for greenhouse gas emissions. So, for any given nation, self-sufficiency must be the goal wherever possible.

This leads to another requirement for countries going forward: one or more relevant leadership positions or other tradable advantages in relation to runaway climate change and its effects. This can be in the form of intellectual property rights, technology leadership, water or grain surpluses, excess renewable energy generation capability, control of strategic minerals – even a willingness and ability to take in immigrants. Each nation must take stock of and re-evaluate its strengths and weaknesses – areas where it has tradable advantages versus areas where it lacks self-sufficiency – on an ongoing basis. (Keeping in mind that the natural tendency is to overrate one's strengths and underestimate one's weaknesses – while in a crisis, strengths can quickly erode, while a seemingly minor chink in one's armour can become fatal.)

   The ultimate advantage that nations can trade on, of course, is military power. "Smash and grab" raids for resources between neighbours may occur, perhaps dressed up under some kind of moral and legal justification. In a nuclear-equipped world, nuclear states have the ultimate defensive advantage, and an offensive threat that only a decent respect to the opinions of mankind and fear of one another will prevent them from taking advantage of.

   The thin veneer of international legality means, in the end, only what nations care to allow it to mean, as shown dramatically in recent years by the George W Bush administration's disregard for the Geneva Conventions protecting prisoners of war. (And by the Obama administration's decision not to prosecute the miscreants.) Nations can depend on protection under international agreements only where backed by the credible threat of force.

   Many countries today have inadequate defence budgets relative to nearby threats, poor military training and little or no experience of war. (Putting troops in the safer regions of Afghanistan or some other failing state, and withdrawing them if they come under fire, doesn't count for much.) These countries may not be in a position to provide their most essential obligation to their citizens – military defence – as things get more and more tense in the decades to come.

   If guns will gain in importance, so will butter. The recent financial crisis should have shown nations the importance of a strong safety net. The state needs to be ready to step in to provide any and all goods and services essential to the population. Our cherished freedoms are quite fragile when anything essential is put under threat, and with climate change, much that is essential will indeed be threatened.

   Nations may need to be ready to move, as smoothly as possible, between degrees of autocracy and democracy in response to fast-changing circumstances. It's ugly and unpleasant to consider what might happen in various emergencies – with the potential for substantial losses of both life and liberty, and the need to preserve each pressing on the other – but it's even more vital to be ready when it does.

Case study: United Kingdom

Britain has many advantages in facing runaway climate change, but has yet to take a true leadership position or put together a coherent plan, even to meet its basic energy needs and the requirements implied by the 2007 Consensus.

However, the UK has recently announced that it will exceed its Kyoto emissions reduction goals for 2100, so not all is in difficulty. The country also has considerable intellectual resources, including the Hadley Centre of the Met Office and others. It's not an accident that it was the UK rather than some other government that commissioned the Stern Review.

Britain is an island nation with a moderate climate. It may even benefit from offsetting tendencies that could keep its temperature fairly close to today's.

Britain is, of course, subject to the planet's overall warming. But warming accumulates over continents; Britain, only a few hundred miles wide along most of its length, is kept cooler by sea breezes. The record 2003 summer high temperatures that caused more than ten thousand deaths in France, for instance, heated the UK a crucial few degrees less.

Also, the UK has always been warmed by the North Atlantic arm of the global thermohaline circulation, which brings warm water heated in the tropics past its coasts. (A good thing, too; the UK is at similar latitudes to Siberia.) This current is in danger of collapsing due to side effects from climate change. If it does collapse suddenly, Britain will be plunged rapidly into cold, which will be a huge shock until warming catches up; but if the thermohaline circulation diminishes gradually as warming proceeds, Britain may have its warming more or less offset by cooling.

As an island, though, Britain faces nasty shocks from rising seas. London is protected by the Thames Barrier, which blocks most sea rises up to a metre. But a combination of ever-rising seas, tidal surges and a badly timed storm or two could overwhelm the Barrier. Rising seas threaten all of Britain's coastlines, with a very high coast-to-land-area ratio meaning everyone will be affected, directly or at a short remove.

Even many inland areas are low-lying and subject, in heavy rains, to flooding. In recent years heavy winter rains have saturated vast areas, teeing them up for flooding, with the actual victims being whichever areas happen to get the last of the spring storms – followed, over the subsequent week or so, by everyone downstream from them.

The same extensive coastlines could be Britain's salvation on the energy front, with extensive channels in its territorial waters that Britain could "mine", so to speak, for tidal energy and offshore wind energy. With good connections to the European grid, Britain could become an energy exporter again, more than replacing its declining North Sea oil and gas reserves with green power.

This is where watching the UK from close up, as I've done as a resident of London for most of this decade, becomes so frustrating. "Failing to plan is planning to fail", the saying goes, and the Brits – artists at muddling through – have so far failed to work energy or climate change issues well.

The country has an incredible history in science and technology, having led the world into the modern age through the work of Newton, Darwin and many others – and having unintentionally led the world into climate change with its "dark satanic mills", the invention of trains, and coal-powered stoves pumping coal dust and CO_2 into the air, creating London's infamous "pea soup" fogs, which were really mostly smog.

But Britain tapped and spent its North Sea oil and natural gas reserves, from the 1970s onward, without ever developing a plan for afterward. No bothering with a petroleum stabilisation fund for the plucky Brits, just a drunken holiday from dependence on foreign energy, followed by a harsh hangover of renewed energy dependence.

Britain, unlike France, failed to create a sustainable nuclear power industry. Like France and, especially, Germany, it became dependent on Russian gas, but failed to develop the storage facilities to buffer itself from disruptions and manipulation. Britain has a ridiculous 10 days' worth of gas storage and is vulnerable to the least hiccup; its continental friends and rivals, closer to their supplier, boast about 100 days of storage each.

Nor has Britain been wise enough to develop good relations with Russia – in marked contrast to Germany's sometimes embarrassing but successful efforts, which have resulted in an offshore gas pipeline directly between Germany and Russia. Something Britain could badly use as well.

Britain is also vulnerable to climate change in a subtle way that the other country I profile here, the US, isn't. Brits are far more attached to their local areas, to local accents and local identities than Americans. Coupled with the nearness of the entire country to the sea, rising tides and flooding will be even more traumatic here than in many other places.

But it's the failure to develop its green power resources that's particularly maddening. In hindsight, green energy from winds and tides could have been funded by, then come on-stream to replace, the North Sea oil and gas bonanza. Nothing like this has happened, and others are grabbing the lead. With no leadership to speak of in green technology or deployment, Britain continues to remain ten to twenty years away from a useful green power capability.

And now the country is entering an energy crisis. It has dithered its way into an imminent shortage of generating capacity, and will soon be rushing to build new coal-powered plants and natural gas-burning facilities to fill in for the declining North Sea oil and gas fields and aging nuclear plants it lacks the skills and political will to replace directly.

New plants are needed to accommodate economic growth, which will in any event be slowed by the brownouts and blackouts that will be required before the new capacity can be completed. (And that will push people from environmentally friendly mass transit such as the electricity-powered Tube, which in past brownouts has experienced service interruptions, into their cars and onto Britain's creaky road network.) Once installed, the new plants will run for decades, mostly on coal imported from South Africa and coal and natural gas from Britain's old friend Russia.

Britain still needs to plan for energy independence and a low-carbon energy future, and it still has strong tides and strong winds all around its islands. A smart grid – which could be among the more cost-efficient in the world, given Britain's high population density – is needed to get power from the best sites to the cities and factories that need it.

Getting popular support, in Britain's system, may not be as difficult as in some other places. The UK has been described as a "Parliamentary dictatorship", with no written Constitution and no Supreme Court to block laws that violate a particular value, no matter how seemingly fundamental. Each party runs for office on a manifesto, and the winner puts it manifesto into law over the following several year period – often, in many particulars, with the cooperation and agreement of the losing party, referred to as the Loyal Opposition.

To many outsiders, British politics seems to be fought on fairly narrow ground. Abortion rights, gay rights, gun control, health care for all and a high overall level of taxation are settled questions – with answers that would mostly be pleasing to a liberal Democrat in the

US. And many international, and even some domestic questions are resolved, at least in part, not at the national level, but by the European Union. So green issues, contentious and resting largely at the national level, should be – but, so far, aren't – crucial in national politics.

Unlike the US, where Democrats have made all the running against vociferous Republican opposition, no important British party yet has seized the green agenda as its own. This has wasted decades of opportunity, but now creates a potential for change to happen quickly. If a green programme were to become part of a winning major party manifesto, it could be passed into law smoothly.

The Labour Government has recently passed a law designed to streamline planning processes, largely by running roughshod over local opposition. Though largely used today to quash opposition to strip mining for coal and the construction of new coal-fired plants, it could be used to speed new green power generation as well.

There is also a smaller, third party, the Liberal Democrats, who are a strong force in local government but who lack a sufficiently clear identity at the national level. And there's a small Green Party which could quickly grow. If election results were sufficiently split among parties, either could be called on to help form a Government – and the "green" planks in their platform swiftly enacted.

People in the UK are increasingly willing to be led. Concern about climate change is quite high in the UK; willingness to pay for change is low, though, and it may take rising seas and continued increases in bad weather and flooding for that to change.

Or perhaps not. Protesters are having some success in slowing new airport runways at Heathrow and Stansted airports. Competitiveness with Germany may also become a factor. The Germans, with a large Green party leading the way, have become world leaders in solar(!) technology, in wind power deployment, and are in the midst of a 20-year programme to improve the insulation of buildings right across the country.

There's also a lot of intellectual and cultural firepower in the UK that has not yet found coherent expression. Many of the leading academics, writers and thinkers on climate change are British – or from other Commonwealth countries, but finding a ready audience (and vital early book-buyers, lecture attendees and so on) in Britain. Brits as a people can accomplish amazing things when they put their minds to something (viz: democracy; ending slavery; industrialisation). If they can put their collective mind to solving the problems of runaway climate change while there's still time to make a difference, Britain can again serve as a model to the world.

Britain has advantages in other important national "musts" as well. It's theoretically easier for an island nation to secure its borders, though a naval presence willing and able to turn back boats full of refugees is needed to enforce this. Britain's EU membership means its borders are open to citizens of other EU members, but this will need to be reconsidered or halted as climate change hits other – especially southern – European countries even harder than Britain.

Planning for rising seas and the subsequent retreat from the coast will be devastating for Britain. The broad outlines of land distribution in England were determined in the Norman Conquest nearly 1000 years ago and then by the Enclosure Acts of the 1700s; the UK is carefully zoned and under various levels of planning restrictions, which are defended down to the last spoonful of earth by various interests. A mere centuries-long global disaster that puts several percent of the country underwater won't easily dislodge this legal, regulatory and cultural briar works.

London is well-protected from small rises but may be impossible to protect against larger ones. The Tube is just one potential early victim of flooding. And most British cities are on the sea. Much of the beautiful British countryside consists of low-lying plains vulnerable to flooding.

So much of the country is exposed to successive tranches of sea level rise that accommodating it will be difficult indeed. There's little that can be done for long-term defence from the sea, so planning is vital. Only about 10% of all the land in Britain is at all densely occupied, so there's room in at least the most literal sense.

Even while suffering this massive disruption, the UK needs to become self-sufficient in food and water. Water is relatively easy, though aqueducts may be needed to funnel water to the drier south – whose chalky soils drain rainwater away too quickly. But estimates are that the UK can only feed about half its current 60 million people – before potentially huge "internal" EU immigration – from its own arable land.

The UK is also part of a corrupt EU fisheries management scheme in which scientists recommend sustainable fishing quotas – and politicians then stupidly increase them by a third or more. This actually suppresses catches by never allowing stocks to recover. The crucial North Atlantic cod fishery, once the source of the fish in millions of helpings of fish and chips, has been wiped out, and is not showing signs of returning. So the seas can't be counted on to help make up the gap in food supplies unless they're given many years of reduced fishing to recover.

Adjusting to rising seas, extreme weather, and possible swings in weather between global warming and possible offsetting cooling may be tricky indeed, putting the UK into food shortages along with the rest of Europe.

By contrast, the Germans – a nationality whose name, in England at least, always seems to have the adjective "bloody" prepended to it – have made the most of a much worst hand. Stuck on an often-frozen plain in north central Europe, cheek by jowl with Poland, France and Italy among others, it's Germany, along with nuclear-powered France, that has helped lead Europe to a stronger stand against climate change.

The UK's strongest current leadership positions with regards to climate change relate to the meteorological expertise of the Hadley Centre, skills in international development, influencing skills within the EU, the Commonwealth and other multi-national and international bodies, and the oratorical skills of its politicians. Only a crash effort to develop and lead in tidal power, along with continued success in cutting emissions, gives the opportunity for a strong leadership position. (Wind is important, but Britain will likely be in a me-too position behind Denmark and the US.)

Britain is in a good position on defence in the literal sense. Its status as an island nation gives it strong defensive advantages, and its armed forces, large and well-equipped by European standards, have seen more combat than any in Europe. Britain's semi-independent nuclear deterrent and long-term strategic relationship with the US mean it can't lightly be attacked, or even pushed around too much.

Finally, Britain needs to strengthen its safety net. Once fairly socialist, Britain now has a Third Way social services system that could harshly be described as European in cost and American in extent. (The NHS and the social housing sector being honourable, albeit incomplete, exceptions.) Rising seas alone mean the Government will need to step up strongly indeed.

Overall, Britain is in a privileged position in relation to runaway climate change, especially in the remaining decades before sea level rises accelerate. Yet the country has done little so far but cope with problems as they worsen – rarely before.

So several scenarios can credibly be put forth – and sadly, the more negative ones can't be ruled out. In the most negative, Britain benefits from its geographical and meteorological good fortune, but little else. It begins by suffering economically from completely avoidable brownouts and blackouts in the years up to 2020. It then suffers both food and fuel poverty, hit hard by short domestic supplies and rising prices for both on international markets.

Then, as warming accelerates, the UK avoids Continental-level droughts and heat waves, but is overwhelmed by waves of European immigrants fleeing these disasters. And a wrenching programme of public works and relocations begins as sea level rises and floods affect more and more of the British coast and countryside.

The country then becomes a giant refugee camp for worse-off countries, especially European ones: muddling through crises, but increasingly sucked into the same disaster as its continental EU partners by wave after wave of their fleeing citizens.

In a more positive scenario, British politics shift rapidly to a green orientation. The country moves to develop its tidal and offshore wind power resources, becoming an exporter of both energy and green technology. It institutionalizes creative and fair solutions to rising seas and flooding.

Closing off unchecked EU and Commonwealth immigration, Britain instead upholds its traditions by selectively taking in immigrants from troubled areas. The country moves toward self-sufficiency in food and water, flexibly adjusting to its uncertain climate outlook. It thus develops an export industry in environmental information and expertise to rival its long-time leadership in financial services.

The better outcome requires early and decisive action, for which it is already quite late. But if such an effort does begin, those who start it may find themselves making friends and influencing people very quickly indeed.

Case study: United States

The US can be viewed in two ways in relation to runaway climate change: like any other nation, as a victim of it; and, uniquely, as the one nation in the world positioned to lead a rapid and maximally effective global response to it.

Perhaps unfortunately for the rest of the world, the US is neither unusually low-lying nor very tropical, so has no large, powerful constituencies who are damaged by the earliest effects and growing impacts of climate change. In fact, with Alaska strategically placed astride the Arctic Circle, the US will be one of the countries poised to benefit from the melting of the North Pole's summer ice cover, though less so than other countries that have arranged to make larger claims.

However, the US has been unlucky in suffering more than one might have expected from some of the general trends associated with climate change. The increased number and severity of the largest storms worldwide has as its poster child the New Orleans disaster caused by Hurricane Katrina in 2005, and exacerbated by the George

W Bush administration's shockingly incompetent and inadequate response. The same Administration that saw New York's Twin Towers fall lost a city to nature, and never saw fit to restore it.

As a side note, climate change naysayers have been quick to assert, and those concerned with climate change all too willing to agree, that the New Orleans disaster was nothing to do with climate change. This is stupid. The city's levees were breached just sufficiently to collapse; in a 0.8°C cooler world, Hurricane Katrina would not have been intensified by a Caribbean warmed by about 1°C, the sea level of the Caribbean would have been about 8" lower, and the levees 2-3" less affected by subsidence, all these factors making them less likely to give way. New Orleans is indeed very likely the first of the world's major cities to fall in a climate change-caused disaster.

The US is vulnerable to further such low probability, high impact disasters at the city and regional levels, but not greatly more so than China or India, for two examples. And, having a random element, such disasters are resistible as goads to comprehensive and necessarily difficult action on climate change.

The American Southwest – including California, with its 33 million people and enormous agricultural productivity, the fast-growing cities of Nevada and Arizona, as well as Colorado and Utah – is another victim of climate change. The entire area has gone into what appears to be long-term drought, buffered for now by Lake Mead, the great reservoir behind Hoover Dam southeast of Las Vegas. (A dam that was built because of the need for public works projects to ameliorate the Great Depression of the 1930s.)

The level of Lake Mead has dropped by about half in recent decades, and may well empty around 2020 – kneecapping agriculture in the region at a time when the amber waves of grain across the Midwest will also be suffering from increased temperatures and more sporadic rainfall. Remaining rainfall and desalination can probably provide individuals' and even businesses' direct water needs, but feeding all these people – and others, in the US and worldwide, who benefit from the region's food exports – will be of critical concern. Yet Lake Mead may need to fall by another quarter of its overall capacity before planners really begin to panic.

The same drought threatening the water supply is contributing to successive waves of wildfires in the region, threatening homes and damaging the environment – even while throwing greenhouse gases into the atmosphere and nibbling away at the total global supply of greenery.

So the US is getting hit surprisingly hard by the early manifestations of climate change. How well positioned is America in relation to the challenges that any country faces, set out at the beginning of this chapter, and which I then analysed in relation to the UK?

The US is certainly in a much better position in terms of its leadership's willingness and ability to face the issues than it was. The US has a new President and administration and a Congress firmly in the hands of the President's party, with all concerned full of piss and vinegar and facing an economic crisis that gives them fairly free rein to tackle climate change as part of economic renewal. And this, to their great credit, they are already beginning to do.

Contrast this to the Bush administration, which blocked international action on the issue, actively suppressed dissenting views and criminally undermined the presentation of scientific findings; or the UK's Labour government under Gordon Brown, who's still as yet unelected, several years after taking office, and prone to mistake bold words for effective action. It may be that only in the US, among major nations, is a U-turn as dramatic as the Bush/Obama reversal likely short of revolution.

Obama, however, is hampered by world-class, but still poor, scientific advice. His notably "green" team of major appointees to scientific, environmental and energy positions – and national security positions also – is well up to date on the 2007 Consensus, but has failed to seriously consider runaway climate change. This is a crucial miss; if any decision-maker in the world could expect their top team to have thought broadly enough to at least construct scenarios around such an important possibility, the American President would be the one. (By contrast, the Pentagon has invasion plans ready to execute for untold dozens of countries.)

So Obama's actions are bound to be insufficiently bold to even lay the groundwork for addressing runaway climate change, due to the failure of imagination and competence by his advisers and by the worldwide scientific community as a whole.

On election, George W Bush led a Republican party strongly opposed to firm action on climate change, and he did even less than promised on it. Today, his party continues to dig in ever deeper as climate change sceptics and even deniers.

Obama and the Democrats look set to take the slightly stronger public support that's now out there for action and run well ahead of it, making energy independence and climate change a "top three" priority when he could easily have gotten by with doing less. It's too bad his campaign commitments and initial, crucial actions are taking place within the playing field defined by the 2007 Consensus rather than a broader framework such as the one presented here.

US Presidential administrations are largely shaped by the campaigns which precede the first term and the actions they take early on, with any second term usually a fairly faint echo of the first. So it will be difficult for a shift toward fighting runaway climate change to occur in what seem likely to be eight years of an Obama administration – and the tendency in American politics is for power to then revert to the opposing party, on current form a frightening prospect indeed, for America and the world.

What about other aspects of a national desiderata for runaway climate change? The US is notable for not having control of its borders, north or south. Patrolling the northern border has not mattered much, except for terrorism concerns – though Canadians may need to worry about an influx of Americans fleeing climate change in the future.

But to the south, several million illegal immigrants have come in over the border with Mexico to meet US labour needs, leading to huge problems even before climate change.

Mexican nationals have recently spread much more widely throughout the US from their former nearly total concentration in the Southwest. This network of countrymen – and a naturalisation programme expected from the Obama administration – will create a pull for more Mexicans into the US, just as many of them are pushed outward from their own country by drought and desertification.

Only some of the most rigid border enforcement and employee verification standards ever implemented in the US – both of which would feel very un-American – will prevent something close to a fusion of the two countries as Mexico and the US Southwest both partly empty north- and eastwards. So a strong border control programme, hopefully accompanied by a strong naturalisation programme for long-time but currently illegal residents – as pushed for by both President Obama and Senator and former Republican Presidential nominee John McCain, among others – is vital for the US, and should be implemented now, while the economic crunch has lessened immigration pressure.

What about planning for rising seas? The US has a huge problem here, with population, economic and political power so concentrated on the coasts that the middle of the country is rather coldly called "the flyover". Most large American cities are on the coasts, and half the state of Florida will be underwater with even a few metres' rise in sea level – rivalled perhaps only by Bangladesh as a large, densely populated area threatened by rising seas. Washington, DC, Manhattan and Silicon Valley are threatened as well.

Even before vulnerable cities are forced to build elaborate defences, or submerged, they'll become more and more vulnerable to storms. Some, especially those on the Gulf Coast such as New Orleans, are likely to be pummelled again and again. Americans are rich, patriotic and generous, but pragmatic, and are likely to adopt a "three strikes and you're out" or similar rule with regards to insurability and federal rebuilding assistance.

So the US will need robust response and resettlement plans, with questions such as military involvement in rescue and rebuilding operations settled well in advance and civil defence drills becoming a norm in some places. Luckily, the US has plenty of land. But it will take a great deal of political will to head off serious problems, including the decades-long knots of litigation to which Americans are so prone. (Think Charles Dickens' *Bleak House*, a desperately sad story of a multi-generational legal battle, as a societal norm.)

Planning for changes in weather will be difficult as well. The United States, as the name implies, is (are) quite strongly federal, which means states and localities have a great deal of power in areas such as land use planning. A "beggar thy neighbour" approach prevails, with many locales competing for development, whereas such issues would be addressed more on a national level in, for instance, the UK.

Incredibly, desert areas of the US Southwest have been the main beggarers in recent decades, with dry cities such as Las Vegas, Nevada and Tucson, Arizona making life easy for developers and growing by many millions of people. Sensible planning has gone by the wayside, with vast acreages of green lawns and lush golf courses scattered about an otherwise desiccated landscape.

The Dust Bowl of the 1930s shows what can happen when rainfall falters in the region, as it's predicted to with climate change. Combined with aquifer and reservoir depletion, and diminished runoff from shrinking glaciers and snowfields, the Southwest and Midwest are strongly threatened. Big problems for these regions could undermine America's entire economy. And the country's preponderant military power rests ultimately on that huge economic base.

Smaller-scale changes in weather will affect the US greatly as well. Government policy has actually encouraged extensive (and sometimes quite expensive) development on marginal land nationwide, for instance through federally guaranteed insurance programmes. "Defend or abandon" decisions will be revisited again and again as rising seas, worsening weather and fire – as well as insurance re-rating and storms of lawsuits – make various areas less and less viable. Without a complete turnaround in planning at every one of America's many levels of government – Federal, state, county or parish, city or town and special districts – disasters will recur, and their after-effects will drag on and on and on.

Remembering that food self-sufficiency is a function of available fresh water, which is becoming scarce, and arable land, which is adequate, self-sufficiency in water and food may be just possible for the US, a drop from its status as a major food exporter, producing about a quarter of the world's food. Even maintaining self-sufficiency is likely to require massive aqueduct and related public works projects, with perhaps some population movement from threatened areas in the bargain.

Food self-sufficiency would be much easier to achieve for the US and Canada combined, a fact which will not have escaped the Department of Defense – an Orwellian name which superseded the old, straightforward one, the Department of War. Nor will it have escaped Ottawa. If the institution of marriage can be rather coldly described as a structured way of selling one's virginity on the best possible terms, the US and Canada may enter some form of marriage in the not too distant future. (Whereas the US and Mexican populations are more likely to be living together, with some of each of them infiltrating into Canada.)

In traditional military terms, the US is uniquely well positioned to "provide for the common defence" of its territory, even if terrorist attacks were to occasionally succeed. But defending all of America's global interests in a world increasingly in turmoil may be impossible. Imperial overstretch is likely just as economic troubles unravel some of the underpinnings of American power.

The period of relative worldwide peace existing for more than sixty years now, since the end of World War II, is often called the Pax Americana. As the sole superpower, the current world order is in many ways America's to preserve or lose. As the reality of, and threat posed by, runaway climate change become clear, the US will face a "use it or lose it" threat to its global leadership position – as in World War II and the subsequent response to Communism.

If it wishes to keep its global leadership position, the US must become the leader in the response to runaway climate change as well. (With the World Wars as a possible model, as others take the heaviest losses before the US steps in to help achieve victory and claim a – perhaps disproportionate – share of the credit.) The US must not simply cooperate with others, and with great effort bring its own emissions down to global standards, but actually lead a halt to and a reversal of runaway climate change.

The American electorate wants something done about climate change, but does not yet seem willing to give up much to get it. So as both candidate and President, Obama has rather cleverly framed most of a fairly aggressive climate change agenda in terms of two closely related goals: energy independence and economic recovery.

This is a good start against surviving, and even laying the groundwork for leadership in, the gradually changing world as seen by the 2007 Consensus. But achieving leadership against runaway climate change, and therefore survival for the US as well as the world as a whole in anything resembling its current political, economic and environmental order, requires four main steps.

In broad outline, they are:

- Achieving energy independence (as the Obama administration has committed to do) with carbon-free power (which the Obama administration has not yet fully specified). The USA must develop leadership in at least one type of green power, with solar power – sited largely in the ever-drier Southwest – powering a national energy grid the obvious candidate.
- Embracing contraction and convergence to first bring its outsize greenhouse gas emissions down to European levels (a 50% cut for the US), then leading advanced nations in driving to zero emissions – and creating enforcement elements relating to trade and aid to keep both developed and developing nations from breaking out of an ever-tightening regime.
- Leading research and development and the creation of a network of scientific, financial, development, relief and compliance bodies, treaty frameworks and other needed elements to stop the progress of, then reverse runaway climate change.
- First researching, then implementing steps to reverse temperature rises and set the stage for restoring pre-warming CO_2 levels, temperatures and patterns of snow and ice distribution – the "Big Chill". It may be better for the US to act alone, or with a "coalition of the willing", than to seek broad international agreement and cooperation from all countries, some of which will actually benefit in important ways from temperature rises of one level or another.

The Obama administration is already well out in front of the American public's tolerance for discomfort on the climate change front, even after sugar-coating the climate change pill with a tasty coating of energy independence rhetoric. So, trying to keep this ambitious agenda within some kind of hailing distance of political reality, it may be that the Obama administration could lay the groundwork for energy independence and a refreshingly cooperative US approach to international agreements in its first term.

Obama could then attempt to buck the trend of relatively quiet second Presidential terms by making "contraction and convergence to zero" part of leading a global war on climate change as part of its major focus for an ambitious second term. This would include beginning to research a Big Chill programme, to be implemented by – bucking the trend again – a subsequent, still-Democratic Administration. ("The Big Chill with Hil", anyone?)

The US will also need to strengthen the weakest social safety net of any of the economically advanced countries, as climate change causes massive disruptions. The Obama administration and the Democratic Congress can lay the groundwork for this in their response to the current economic downturn and with a new health care framework.

The alternatives for the US to a correct and comprehensive approach to runaway climate change are stark. America could lose the ability to feed itself, face international opprobrium for its outsize emissions and be dictated to by an ever-less-snowbound Canada, a rising China, a resurgent Russia and a European Union leading the economically advanced nations' response to climate change – all while facing millions or even tens of millions of illegal immigrants and internal refugees. The American Century could fade to a distant memory, and even the American experiment find itself threatened, not least by its own gun-toting citizenry.

Alternatively, strong leadership by the US in the face of runaway climate change could allow the US to quickly go "from worst to first" on the issue. If it also helped tackle the other, related global challenges referred to in this book, the US might build a renewed base for economic and political pre-eminence. The first American Century might be followed by another – perhaps miserable, for many people in America and around the world, but not the unmitigated disaster it's looking to be from here.

Appendix A. FAQ

Q. What does "runaway climate change" mean?
A. There isn't an official definition, but the term is used to refer to global warming – initially caused by humanity's greenhouse gas emissions – setting off changes in the environment. These changes lead to more warming, which leads to other changes, which lead to more warming – independent of emissions, and beyond the control of humans.

But how can it be possible for warming to take off on its own?

Chemical and biological processes remove greenhouse gases from the environment and store them. Some of the ways in which carbon is stored render it less accessible to release, such as carbon dissolved in the seas, and others allow it to be more easily be released back into the environment, such as absorption into plant life.

In addition, ice and snow reflect sunlight back into space, keeping the Earth cooler. Though not directly involving carbon, snow and ice cover is still referred to as a carbon sink.

Runaway climate change is when these carbon sinks stop contributing cooling to the environment, stop absorbing carbon from the environment, and/or start emitting carbon back into the atmosphere or oceans at a fast enough rate that they cause further warming. "Runaway" means that they "tip" each other into further changes, without the need for further input from human-emitted greenhouse gases.

I estimate that climate change became runaway climate change in about 1983. The North Pole's ice cap seems to have started melting in about 1970, and had lost half its mass by about 2007, so it was about one-sixth melted – at a statistical "tipping point" – in about 1983. Further research would be needed to fine-tune this estimate.

Now that the North Pole's ice cap is doomed, extensive permafrost deposits in the North are likely to be as well, which would be disastrous. Indeed, they seem to be starting to tip today – and are then likely to be followed, more or less rapidly, by other carbon sinks.

Q: What can I say to people who still believe there is no global warming?
A: Simple; after Galileo: "And still, it melts". Ask them what could have caused the steady, 100-year-long shrinkage of glaciers worldwide, including glaciers in tropical and temperate zones, except truly global warming? The loss of glaciers alone is decisive evidence of climate change. And only a reversal – not a plateauing, but an actual turnaround – of snow, ice and glacier loss into gains, extending for a decade or more, would be enough to show that global warming had substantially reversed itself.

The melting of the North Pole's ice cover is also prima facie evidence, but perhaps less convincing to a layperson than more accessible, and more easily observed, temperate zone glaciers.

Additional simple, convincing evidence is the lengthening of growing seasons worldwide by about a week and the movement into temperate areas of tropical plant and animal species, such as malaria-carrying mosquitoes globally and Lyme disease-carrying ticks in the US, plus unprecedented levels of drought in south-eastern Australia and the US Southwest. Those who question warming must explain these changes – and the lack of changes in the other direction – as well. (They won't be able to.)

Q: What about those who claim warming is real, but not caused by people?
A: It seems quite clear to many that as great an effect as warming of the entire planet couldn't be caused by people, who seem so insignificant compared to the planet. This is more an emotional than a logical argument, based on what feels right or seems sensible rather than careful study. It's unlikely you can fully convince the holder of such a belief by reason alone; it will take dramatic events and social pressure to do that.

Logically, there are two answers to this argument. The first is that, with the CFCs problem and the resulting hole in the ozone, we have indeed demonstrated that we can cause just this kind of global change – and that, by reversing the pollution that was causing it early enough, we can reverse the change as well.

The second answer is that, even if global warming was caused by, for instance, increases in solar radiation, we would still need to do everything urged in this book. We would still urgently need to stop depending on oil because we're running out, and on coal because it's so polluting, in particular with its soot damaging the cryosphere. And we would need to nearly eliminate greenhouse gas generation in any event to stop the acidification of the seas, which threatens a major part of humanity's food supply.

Additionally, we would need to start blocking some of that excess solar radiation, fast, through some kind of geo-engineering. So the same responses (and then some) are needed even in the unlikely event that an otherwise undetected increase in solar radiation is the cause of global warming. It would not in any way be a rationale for complacency, though that's generally how it's used.

Q: Isn't calling climate change "runaway" defeatist?
A: There are two competing goods here. One is to get the science right, because how else can we battle the problem? The other is to keep a positive attitude, because, again, how else can we battle the problem?

The problem comes when being upbeat in approach is held out as a reason to modify what one says – or even what one allows oneself to believe – about the true state of play. This is just an optimistic version of what climate change deniers are accused of: coolly manipulating scientific results to suit political or economic ends.

Describing climate change as "runaway", however accurately, does at first glance seem to invite defeatism. Humanity has so far proved incapable of cutting greenhouse gas emissions or stopping deforestation, even with good reasons to do both. Nor have we come up with technological solutions for removing carbon pollution from the environment. Saying that we di indeed have to do all of these, and then some, might seem overwhelming – and enervating. But if climate change really is runaway, as I assert throughout this book, then only an appropriately large response can master it. Continuing to hide our heads in the sand will hurt, not help.

Q. When should the term "global warming" be used vs. "climate change"?
A. Global warming is the more descriptive term for the overall process – the globe, as a whole, is warming. However, it doesn't capture local variations that can see regional stasis or even temperature decreases localised in extent or in time. Nor does it capture the increase in extreme weather events and the shifting of the world's climactic zones due to warming. So climate change is the preferred, more inclusive term among those who study the phenomenon closely.

However, a recent Google search count showed nearly a ten to one preponderance of searches for "global warming" over "climate change", so to communicate effectively, the term global warming must be used as well.

Q: Why are you so critical of what you call the 2007 Consensus on climate change?
A: The 2007 Consensus reflects a lot of good and important work, and its effective communication to a worldwide audience was an amazing accomplishment. However, there are large faults and gaps in the findings that call out for criticism, in order to build on stronger foundations going forward.

To clear the space required on which to build toward a fuller understanding of humanity's situation, I need to point out ways in which the 2007 Consensus is in error. This process is called "problematisation" by academics, in the sense that one has to show there's a problem before offering a new understanding as a solution. This process often requires offending someone, or several someones, but all involved are grown-ups; sooner or later, as research results accumulate, they'll take any new findings on board and move forward.

Q: Why are most conservatives, especially in the US, so opposed to mainstream scientific thinking on climate change and the need for action?
A: US conservatives in particular have developed a habit of reflexively opposing government action – except military action, ironically the most complex and easily mismanaged type of action government undertakes – so oppose and deride any chain of reasoning that might seem to require extensive non-military government action. Used as a one size fits all rule, this is, of course, stupid; but enough government actions are insufficient, inefficient or just plain wrong that conservatives always have plenty of opportunity to apply the "stupid" tag as well.

Conservatives claim to be in favour of preserving the best in today's world for future generations, and admit to being the party of big business. The business part is certainly trumping the responsibility part among America's conservatives. The commonly made claim that future people, being richer, will be better placed to deal with the resulting problems is a breathtakingly unconservative approach.

The third key element is a tendency, again strong among American conservatives, to put themselves in opposition to science. This is partly because of the strong influence of religious conservatives in the overall movement, for whom the theory of evolution is a particular bugaboo, and partly because of the populist appeal of criticizing scientific claims. Conservationists who try to explain the need to save particular endangered species are among those easily lampooned by conservatives.

Those who believe that climate change is occurring are part of the problem as well. As a group, supporters of action keep moving the goal posts on how much reduction is needed (due to having taken an engineering rather than a principles-based approach to the problem); initial solutions such as first-generation biofuels and ineffective, environmentally hazardous "green" light bulbs have caused many problems of their own. And supporters keep saying that fighting emissions will be cheap, ignoring the true, global scale of the problem and the real suffering that will be caused to many along the way. Nobody's perfect, but those of us on the side of taking action have hardly covered ourselves in glory so far.

We will need the active cooperation and involvement of nearly everyone to solve the problem of runaway climate change. For one example, there were fascists and sympathisers in the UK right up to the start of World War II, and large numbers of isolationists in the US right up until Pearl Harbor, nearly all of whom then joined heartily in the fight against fascism. Those of us who already favour strong action must avoid making bitter enemies today of people whose cooperation we will badly need tomorrow. The term "climate change deniers", purposefully reminiscent of the damning "Holocaust deniers", is one example of language that should be used sparingly, particularly when applied to individuals.

Q: Why are many businesses in favour of action on climate change, even if it costs them dearly?
A: Most businesses hate uncertainty. When a problem becomes too large to ignore, but a solution is not in place, business must plan for an impossibly wide range of possible problems and potential remedial actions. At such a point, businesses tend to favour even imperfect action over uncertainty. Climate change is at this juncture.

Having said this, some praiseworthy business response to climate change, such as the Plan A programme from the UK's Marks & Spencer or the "first carbon-neutral bank" status achieved by global banking giant HSBC, will not necessarily lead to follow-on action by their competitors. Businesses love to differentiate themselves, and only the first tranche to respond to a new issue really achieve differentiation. Other businesses need additional strong reasons to act in a way that will, after all, appear to be merely following the leaders. That's why I strongly favour governments, businesses and individuals moving toward establishing their own "green supply chains", which can make strong action by all businesses a must, rather than an option.

Q: What will it take to get out of the trouble we're in?
A: The needed goal can be summed up in a three-letter acronym: SQA. This is short for "status quo ante", or "the way things were previously". We need to return to pre-warming levels of greenhouse gases in the air, acidity in the seas, extent of forests to absorb and hold carbon, variety and number of plants and animals in the environment, and extent of the cryosphere.

In order to get there, we need to use contraction and convergence to zero as our primary guideline in reducing greenhouse gas emissions to pre-warming levels; to use geo-engineering, beginning with reforestation, to clean up the excess CO_2 in the air and seas; and to implement a "Big Chill", a decades-long period of slightly sub-normal temperatures, to restore the cryosphere.

How much time do we have to do this? It's too late to prevent runaway climate change, so now we have to fight and reverse it. It's basically a race between increasing death and disruption – in the human and natural worlds – caused by climate change and by other pent-up problems, as described in this book, versus the human capacity to implement the needed extensive programme of remedial action. Loss of our accustomed climate and of other species will fairly quickly reduce the carrying capacity of humans by Earth, so we must begin to act very soon to have even a fighting chance of avoiding the most serious consequences.

Q: Will fighting runaway climate change be expensive?
A: I compare the fight against runaway climate change to a world war. It will be very expensive. Yet not fighting is not an option – and, as a side effect, the battle may ultimately result in a brighter economic future than if the war had never taken place.

Large-scale infrastructure such as power plants is not normally replaced until it's so irredeemably worn out as to be dangerous – and it's normally replaced with something as similar as possible, as nearby as possible. Only during wartime is such infrastructure so comprehensively destroyed that people really rethink what goes where. And at that point both winning and losing countries are so broke that they usually do the cheapest thing they can.

So to replace not-yet-depleted carbon-emitting infrastructure such as coal plants ahead of schedule, to rapidly replace trains and trucks and cars that run on fossil fuels with electric versions, to redesign global transport networks to minimize air travel and so on will be very difficult and expensive and meet tremendous resistance.

There's also tremendous opportunity cost in all this. Tying up brainpower, capital, planning capability, construction capability and so on to replace functioning equipment with other functioning equipment, instead of innovating elsewhere in the economy, means losing out on many potential gains.

To motivate the needed effort, the single most important step is to establish a price for carbon, to encourage conservation and development of green energy sources. This is going to hurt a lot, including pain to those who can least afford it. The only consolation is that it will hurt far less than letting energy prices skyrocket and fossil fuels run out with no replacement , which is simply unacceptable. Yet it will be very brave governments indeed that try to make this case to their people. (With "antis", such as the current US Republican leadership, cynically baying at their heels.)

It may well be that fighting climate change turns out to be cheaper than we can imagine now, as many other steps to cut emissions of various pollutants have been – but some aspects may be more expensive than expected, as, in particular, many pollution cleanup efforts have been. Whatever the net result, "nearly free" is not an option from here, except perhaps as a net result in the very long term.

Q: What's the single most important factor in successfully battling climate change?
A: The single biggest problem for future generations in battling climate change is the diffuse and multinational nature of the problem. From a national point of view, climate change causes are separated from effects, and emitters are often not the ones who suffer most, neither by geography nor across generations. Yet countries continue to be the only consistently effective vehicle for garnering consent and maintaining economic and political control, and only in the short term do we have the opportunity to head off disaster.

The only entity that can effectively lead countries to battle climate change is the economically and militarily largest country, the United States, which of course is also the leading cumulative polluter. The US must lead in setting up and repurposing international organisations for the fight; in creating treaty frameworks to promote action and punish inaction; and in financing and rewarding action against climate change through grants, loans and trade terms. In many respects, it's the fight against communism all over again. The US will be the world's leading nation through this century, as it was in the last one, only if it meets this new challenge head-on.

If the US fails to fully take up this challenge soon, not only America's but the world's opportunity will be lost, and we'll be headed for Hot Earth – a hot world with extensive deserts.

Q: This book recommends geo-engineering – technological fixes – along with dramatic reductions in greenhouse gas emissions and an end to deforestation. But why bother with painful cuts in emissions if we can engineer the problem away?
A: There are two answers to this. The first is that we don't have geo-engineering fixes yet. Any given such fix is likely to turn out to be nearly or completely unworkable, damaging and/or very expensive. So we have to proceed as if any such solutions will be of limited help, or not available when needed, or too expensive to depend on much.

Also, technological solutions are partial. Techniques which counteract warming don't stop acidification of the seas; taking CO_2 out of either the air or water helps one but not the other. They're also controversial, and international consensus should in theory be gained before using any of them. And they don't directly address Peak Oil nor, though the date for this is still off in the future and uncertain, Peak Coal.

Even if good technological fixes are found, they'll be expensive. Their cost will likely be added to the cost of the practices they counteract, slowing their adoption. So we need to cut emissions even as we pursue technological fixes.

I liken our situation to World War II, in which the Nazis and the US both energetically pursued doomsday weapons – yet the course of the war in Europe was decided by technologies available when it began. The US finally developed the atomic bomb after the war was over in Europe and almost decided in the Pacific, and only two bombs were built in time to be used. They did make a difference in finishing the war with Japan sooner, but also opened the door to immense problems that are with us still. Technological fixes may play a similar role in fighting runaway climate change.

Q: How dare you, as a non-scientist, introduce what you call a model for runaway climate change?
A: Simply because it needed to be done for my purposes in this book. The possibility of runaway climate change has been implicit since global warming due to greenhouse gas emissions was defined by Arrhenius more than 100 years ago. And it's been explicit for many years, nowhere more clearly than in Mark Lynas' remarkable recent book, *Six Degrees* (2007), which spelled out one sequence by which it might unfold.

I admit it's pretty brassy of a non-scientist to step in and take on what needs to be, in the end, a scientific effort. I can only hope some actual scientists are already, or will soon begin, filling in some model resembling the one given here, as no topic on Earth is more important.

Q: What will happen if we fail in the battle against runaway climate change?
A: This needs a great deal of study, but the ultimate destination seems to be the next major stable state, in the warmer direction, found in past versions of Earth's environment: a world with Jurassic-type conditions – but, for a long time, relatively lifeless, one-third covered by desert – temperatures averaging 10°C or more above today's, no permanent ice on any part of the surface, sea levels perhaps 120m higher than today's, and much greater greenhouse gas levels in the atmosphere and the seas. I call this Hot Earth.

In many millions of years, after life has a chance to evolve to fit the new conditions, there isn't any reason that this version of Earth couldn't become as biologically productive and even as liveable for a large number of humans – at different latitudes – as the current one. The problem is the very long expanse of time between here and there.

Over the next few centuries, weather will be erratic and undependable as we move from our recent climate toward the new one, and sea levels will be steadily rising by several metres per century. Human populations will need to move repeatedly as seas rise and habitable zones fundamentally shift, masked by the froth of erratic weather. And the "age of loneliness" – as the upcoming period, after so many of the familiar species on land and sea have died out, has been called – will be a tough time to make a living, with many of the services provided by our current environment reduced or eliminated.

Human populations are likely to drop sharply, with our very survival at risk from the sharply reduced carrying capacity of the severely damaged environment and from probable infighting among our survivors. We can only hope they don't use nuclear weapons.

Perhaps the humans that remain will find a tenable way to survive in the barren transitional environment. Over millions of years, barring human interference, new life forms will evolve that will thrive in the new conditions, which may come to resemble those in the movie Jurassic Park – though what the dominant large land and sea creatures will be at that point, no one can know.

Perhaps pigs will outlive most other animals, current porcine species will radiate to fill many of the available ecological niches, and we'll have Jurassic Pork.

Q: Why are you against GM crops?
A: My reasons are a bit subtle, and subtle logic may not withstand the pressure of the emergency we're entering into; but perhaps I can at least bring up some dangers to which all of us should be alert. The three reasons that convince me to oppose genetically modified (GM) crops are the increasing impact of water shortages on food production; the quality of food; and the future of life on Earth.

Water shortages are increasingly impacting on food production, with every kilogram of grain representing approximately 1000 litres of water. GM crops can be used to grow drugs, not just food – a phenomenon called biopharma. Biopharma, along with biofuels, being highly profitable, divert the best, and best-watered, land away from food production.

More subtly, the quality of the food we eat – as measured by its nutritional content, as well as its increasing burden of poisons (for no insecticide can be anything else) – is gradually declining. Much farmland has lost the nutrients that supported the crops grown on it, so that now even weeds won't grow on it when it hasn't been newly fertilised. Each season a crop is planted, the land is sparsely nourished with a load of fertiliser, insecticides and weed killer are sprayed, and the crop is harvested. In many cases the resulting food has a 25% or greater reduction in measured nutrients from crops of some decades ago; processing then takes out more nutrients, all of which hurts our health.

GM crops will be deployed against both of these arguments, with claims they can be made to use less water and to get higher yields – measured by weight, not nutritional value – out of marginal lands. What's more certain is that they'll be used to force-fit familiar crops to inappropriate conditions to meet immediate market demands, rather than switching to different cereals, for instance, that are already, and more fully, well-adapted by evolution to the same difficult conditions. (Which would require re-educating the markets, new inspection regimes, new recipes and so on – all complicating, and profit-reducing, steps in the short run.)

Artificially developed plants are more likely to be vulnerable to pests and disease. They'll provide profits to the corporations that control them at the expense of the rest of us. They're also likely to exacerbate the trend to growing unhealthful food on poor soils by the use of lots of energy-intensive fertiliser and scarce water.

This fits into my final concern – the future of life on Earth. Traditional farming and stock raising were better integrated with the natural environment around them, subject to the same or similar constraints of weather, water and soil quality as natural ecological niches, and with less abrupt transitions between the agricultural and natural environments.

The farther we go down the GM path, the more we impact the natural environment. Not only are the farm and the feedlot ever less natural, to the likely detriment of the environment and our health; we'll be tempted to push GM plants and, soon, animals into the natural environment itself. Imagine seas stocked with little but frankenfish after the originals fail to survive our current onslaught against them.

Artificial genes are likely to lack the flexibility and redundancy of genes that have evolved through billions of years of pressures across all sorts of climatological and competitive environments. (The current biological depth, and repeated examples of recoveries across billions of years, are part of the reason we can count on an unengineered biosphere to eventually rebound even from Hot Earth conditions.) Contamination of the biological environment by modified genes could harm the capability of life to fully regenerate itself, even on timescales in the millions of years, as changing conditions expose new inadequacies in the engineered genes.

In addition to the specific reasons outlined above, I have a healthy respect for humanity's ability to screw things up, though we also get very many things right. Genetics are so fundamental to the natural world, including us, that delay and further study may be the wiser choice.

Europe's current rejection of GM crops might not survive the pressures of our upcoming environmental crises; but it might be a good thing for the health of Europeans, and perhaps for the long-term health of the rest of the world as well, if it did.

Q: Where is the Obama administration on all this?
A: President Obama himself is well out ahead of most other leaders and the American public on climate change. The Obama administration has some of the top leaders on climate change in most of the right positions to make a difference, but they haven't yet been able to see out of the box imposed by the 2007 Consensus.

Even if the 2007 Consensus were a true picture, rather than understating the problem, Obama would risk running out of time before garnering the support needed to effectively respond to the crisis the 2007 Consensus describes, let alone anything worse.

To go even further and see that climate change has gone runaway, and needs an even greater response, Obama – a lawyer by training, which is sort of the opposite of a scientist – would need a lot of help from the experts around him and from the worldwide scientific community as a whole; help that is unlikely to be forthcoming in his first term and only with luck in time for a possible second term.

Appendix B. Rallying Cries

Can the message of a complex book like this one be summed up in a few short phrases?

I think it can – here follows my attempt to do so:
- **Stabilise at 280ppm**. The level of CO_2 in the atmosphere before industrialisation began was approximately 280ppm. The current tipping of carbon sinks shows that the climate is quite sensitive to higher levels of CO_2, so for safety's sake we must return the atmospheric CO_2 level to 280ppm as soon as possible.
- **Return to pH 8.2**. The seas, having increased in acidity 30%, from a pH of 8.2 to 8.1, are on track to reach a deadly pH 7.8 by mid-century. The current acidity increase must be reversed, pollution decreased and fisheries protected to restore their health and productivity. This isn't green namby-pambyism – not an effort to "save the snail darter" at humanity's expense – but just simple and necessary steps to avoid (further) disaster and to help feed a growing worldwide population.
- **Save the rainforest**. Deforestation must be stopped immediately, which will be a wrenching, worldwide effort. Reforestation must be begun, with the goal of returning the Earth's forested areas to close to their former extent, about double what they are today.
- **Save the cryosphere**. The cryosphere is everything frozen on Earth – the North Pole's ice cap, the Greenland and Antarctic glaciers, temperate zone glaciers and even snow cover. All of it is at risk from runaway climate change. We need a return to pre-warming snow and ice cover to restore the Earth's heat balance, as well as to meet key human needs such as irrigation from snowmelt. Tropical and temperate zone glaciers are the "canary in the coal mine" that have been the first clear signal of runaway climate change.
- **Status quo ante (SQA)**. This Latin phrase, commonly used in lawsuits, means "the previous condition". We need to return the Earth to a pre-warming equilibrium among temperatures and atmospheric CO_2 levels; ice cover and sea levels; non-acidity of the seas and forest cover. This will restore the environmental balance in which human civilisation developed and grew. If we

don't restore and maintain this balance, the Earth will shift to a different, much hotter equilibrium far less friendly to humans and to the species that make up today's natural environment.
- **Clean up the CO_2**. One of the key mistakes of today's climate science advocates is believing that we can leave CO_2 in the environment and wait for (damaged) environmental systems to flush it out. We have to directly remove or neutralize CO_2 out of the atmosphere and seas to buy time for emissions reductions, to stop warming sooner and to save marine food chains.
- **The Big Chill**. In order to return to the pre-warming status quo, we need to restore the North Pole's ice cover and temperate zone glaciers. This will require a period of temperatures below pre-warming norms – perhaps 1°C to 2°C below for several decades. To do this will require near-zero CO_2 emissions plus one or more geo-engineering efforts to actively lower the temperature.
- **+2°C is unjust**. One tenet of the 2007 Consensus on climate change, as represented by Al Gore's book and movie, the 2007 IPCC Report, the Stern Review and other sources, is that global average temperature increases must be held to +2°C or runaway change might begin. But even if it were possible, holding at a temperature rise of +2°C would cause water shortages, crop failures and sea level rises that would damage the prospects of and dispossess millions, mostly in poorer, crowded equatorial areas. It would be hugely unjust for the rich of the world to inflict this on the poor. (Unless they're prepared to issue the requisite visas, transport vouchers and resettlement money to bring them into the rich world, as New Zealand has helped do for the tiny island nation of Tuvalu.)
- **Holding at +2°C is impossible**. Even if it were morally defensible to allow a 2°C rise in temperatures, we actually tipped into runaway climate change about 20 years ago, when warming was at about 0.3°C. We must stop greenhouse gas emissions on an emergency basis and start work to reverse warming now; waiting until we reach +2°C will see us well launched toward Hot Earth conditions, with temperatures rising 10°C or more and extensive desertification.
- **No Hot Earth**. Runaway climate change, which has already begun, will take the Earth's climate to a new equilibrium unless reversed. The most likely new equilibrium state features temperature increases of 10°C or more, deserts covering a third of the land, greenhouse gas levels triple today's, no ice cover, sea levels about 120 metres higher and a far less rich and diverse natural environment – Hot Earth.

Appendix C. Selected Resources

For links to the resources in this chapter and throughout this book, visit getridofwarming.org

For all the complexity of the topic, catching up on current mainstream thinking around climate change is surprisingly easy. (A summary of the mainstream view is presented in this book, but it's unfair to its proponents for you to get their viewpoint only as reported by a critic.)

All you need to do is read Al Gore's best-selling book, *An Inconvenient Truth*; optionally, see Al Gore's Academy Award-winning documentary, also titled *An Inconvenient Truth*; then visit Joe Romm's Climate Progress, a most useful and accessible blog on climate change, and read a representative sampling of his recent posts.

In about six hours, and at a cost of less than $40 – or close to free, if your local library has the Al Gore book and movie, plus Web access – you can be better informed on climate change than the vast majority of people worldwide.

If you also read George Monbiot's *Heat* for an individually focussed account, and Bjørn Lomborg's *The Sceptical Environmentalist* for one type of opposing view, as well as having read this book for another, you'll be the next thing to an expert.

Use the references here, as well as other resources they point to, to bring – and keep – yourself up to date.

Books and films

The books below are the few I consider essential; there are many other books, reports and papers worth reading, but these are the core few. In my view, however, all are flawed by putting the onset of runaway climate change off in some still-avoidable future. (Lynas' book bridges the gap by spelling out in some detail what the beginnings of runaway climate change would be like.)

All of the following are available on amazon.com and through bookstores, except where otherwise noted. Events relating to climate change are changing so fast that any book begins to become outdated before it reaches the newsstands. Fortunately, you can use the recommended Web sites listed below to bridge the gap.

An Inconvenient Truth by Al Gore – book and movie

The best single book on climate change. Though a full-size book (and an expensive one), it's packed with pictures and graphics, so the actual text is quite short – making it a quicker read even than this book.

An Inconvenient Truth is a must-read for anyone wanting to understand and discuss current thinking and action on climate change. It's also, in a pinch, everything you need. It even, as I pointed out in Chapter 3, anticipates the central argument of this book – only, unfortunately, to dodge the question at the end.

Taking Gore's book on its own terms, as a brief, accessible yet authoritative introduction to climate change, there are only two major faults I'm aware of. First is that it dramatically understates emissions from China and the likely growth of developing world emissions. Second, and not unrelatedly, it's overoptimistic and dangerously, in my view, wrong in asserting that climate change can still be stopped simply by gradually reining in the human actions contributing to it.

The book is accompanied by a documentary movie of the same name that's a great resource in its own right – and even stronger in implying that climate change is already runaway – but not, to my mind, a substitute for the book.

The book and movie, along with Gore's live presentations, changed the world, making this book and much else possible. You need to read Gore's book and, for a complete view, see the movie to understand the basis on which current informed discussion of climate change occurs. That, plus visiting one or more of the Web sites listed below.

Heat by George Monbiot (2006)

This is not only one of the best books on climate change, as attested by its sales and public and critical acclaim, I think it's one of the best books ever. George Monbiot is a famous British environmentalist whose views in other fora sometimes approach being extreme. But in this book he simply focuses, one topic at a time, on the changes he would have to make to reduce his own carbon footprint by 80%.

The result is funny, entertaining and thoroughly enlightening. Read this book to gain a deeper understanding of the points raised in a more general way by Gore, and to educate as well as to amuse yourself.

The Stern Review of the Economics of Climate Change **by Sir (now Lord) Nicholas Stern, et al (2007)**

One of the most frustrating books I've ever read. Stern brings together a vast amount of information and analysis and presents most of it – with the exception of some econometric modelling – in a relatively understandable and easy to read format. But he blithely accepts key information he should have known better on – most importantly, that China's emissions are both low and controllable by similar means to those of the West. (Stern is former Chief Economist of the World Bank, for crying out loud.)

Stern also leaves out any value for human life beyond its economic contribution; which, as George Monbiot pointed out in an essay in response, leads to chilling conclusions. A serious work, but one with serious flaws.

IPCC Fourth Assessment Report **by the Intergovernmental Panel on Climate Change (2007)**

Widely available for purchase. Also available for free download (PDF files) at:
>http://www.ipcc.ch/ipccreports/assessments-reports.htm

Summary for Policymakers:
>http://www.ipcc.ch/press

Emissions Scenarios (2000):
>http://www.ipcc.ch/ipccreports/special-reports.htm

This Report and preceding Reports, and the Emissions Scenarios from 2000 used in the 2001 and 2007 Reports, make up most of the basis of the world's current understanding of climate change. Almost every book, article, broadcast story, presentation or discussion about climate change accepts and is based on them, directly or indirectly. And they are seriously flawed, as described in Chapter 3.

The Scenarios are baseless – you can verify this yourself by reading them, or at least the Summary in the Scenarios report; the Fourth Report's Summary for Policymakers is so confusing as to seem deliberately so (including in its heavy dependence on the Scenarios); temperature increase predictions ignore rapidly increasing emissions from industrialising countries; seemingly precise sea level rise predictions exclude the major component of future sea level rises, ice melting; and the whole is unwontedly optimistic in its predictions. A few IPCC scientists have split off into separate projects or begun laying the groundwork for a much more engaged Fifth Report, due in 2014. This is badly needed; time is running out for the IPCC and for the world.

Six Degrees by Mark Lynas (2007)

Mark Lynas is a real hero of the fight to save the Earth, but I found this book a bit hard to get through. It's organised by the consequences of warming, one degree at a time; but since the average we all refer to means very different things on sea, coast and interior, at the equator vs. the poles, and so on, the degree-at-a-time view is not that effective as a framework for a whole book. But Lynas writes knowledgeably and with real passion.

Plan B 3.0: Mobilizing to Save Civilization by Lester Brown (2009)

Widely available for purchase. Also available for free download (PDF file) at:
> http://www.earth-policy.org/Books/PB3/Contents.htm

This is a new, updated edition of *Plan B*, a wise and humane book that takes on the whole problem: not just climate change but water shortages, overpopulation and human development. The fact that the book is available for free download is amazing and helpful. Even if you've purchased the printed book, the PDF is useful for searching, excerpting and sharing key points. There is some "layering" visible of facts given and arguments made among different editions of the book, but this is a vital and important work.

Seven Years to Save the Planet – The Questions and Answers by Bill McGuire (2008)

This book is not a major contribution in the same way as those listed above, but it's relatively up to date and easy to read. As with the other books, the author sets out as his premise that there's still time to save the planet by reducing emissions and stopping deforestation – then sets out facts that go far toward showing that there isn't. The press has gone relatively quiet about climate change since 2007; this book proves that it's not because things have gotten better.

The Skeptical Environmentalist by Bjørn Lomborg (2001)

This is a carefully thought-through book that questions climate science's arguments and undermines, without wholly refuting, some of them, then argues for fobbing the whole thing off on our descendants – who, being richer than us, and knowing rather than guesstimating what the impacts will be, will be ready and able to buy off the forces of nature.

In the movie *Broadcast News*, it was said of the Devil that, if he showed up on Earth today, he wouldn't advertise himself by sporting horns and a tail, nor by being visibly evil; he'd just make everything "a little bit worse". If this comparison, and a queasy feeling in your stomach, strike you as you read Lomborg's seemingly reasonable yet subtly misleading arguments, you won't be the first.

This book is the best single resource for understanding the "anti" argument to the 2007 Consensus. To many of us seriously concerned about humanity's future, it's just this side of respectable, given what was known at the time it was written. The author has, less defensibly, followed up with a new and more stridently titled book, *Cool It* (2008), that makes a similar argument.

The Guide to Low Carbon Lifestyles by Mukti Mitchell (2007)

Available for free download (PDF file) at:
    http://www.lowcarbonlifestyle.org/guide.html

Mukti Mitchell is a positive and upbeat British green activist who promotes low carbon lifestyles. His most dramatic effort so far, a microyacht tour around Britain, received the support of royalty, top politicians, leading businesspeople and others.

The Guide to Low Carbon Lifestyles avoids hype and bombast (ahem) and presents, simply and accessibly, a guide to choices that will reduce your carbon impact. It can be used as a companion to the theoretical and long-term approach in this book. Though the carbon emission figures and the (relatively few) cost figures given are UK-specific, they're easy to re-interpret for other countries, other climates and other currencies, and the relative costs and benefits of different actions and inactions are clear.

Web sites

There are a fair number of good sites about climate change – and a lot of bad sites and bad writing on both sides of the issue. This is largely due to tension between different schools of thought – the sceptics and deniers in one camp, the climate science advocates in another, plus the new strand represented here. ("Runaway climate science advocates", perhaps?)

The combination of real differences of opinion with the tendency towards bombast and hyperbole found on the Web generates enormous amounts of heat and much less light. Even the mainstream media joins in; one can find articles stating the end is nigh, and others saying that it's all a hoax, often in the same newspaper or news Web site on the same day.

Yet the instantly updated nature of Web news and blogs makes the Web a vital tool for keeping up. The very short list of sites below should be a good starting point.

The Institution of Mechanical Engineers report

Web address:
  www.imeche.org
Report available for free download (PDF file) at:
  http://www.imeche.org/about/keythemes/environment/
  Climate+Change/Adaptation/Adaptation+Report.htm

The Institution of Mechanical Engineers, also known as IMechE, is a British chartered royal society that has produced a landmark report about the impacts of climate change. The report, called *Climate Change: Adapting to the Inevitable?*, anticipates several of the arguments in this book, and is referred to in Chapter 4. The IMechE Web site has the report and selected press response.

Real Climate

Web address:
  www.realclimate.org

Real Climate is the mothership of climate blogs, with comprehensible postings on a range of topics from real scientists. The site hosts a steady stream of updates on topical issues relating to climate change, with constant reference back to the scientific underpinnings of the subject. The one "must read" for those following climate change.

Climate Progress

Web address:
  www.climateprogress.org

Written by Joe Romm, a former US Energy Department official, Climate Progress keeps its many followers up to date on major and some minor news on climate change, maintaining a good balance among news, science and politics. Romm takes positions on the issues – he is, for instance, largely against geo-engineering, as I would be as well if I thought we could survive without it – but explains and cites evidence supporting his stances as he goes along. Climate Progress points to many other sites, both in agreement with and disagreeing with Romm's points.

Guardian Environment

Web address:
  www.guardian.co.uk/environment

The Guardian newspaper's environment news and opinion site is perhaps the leading such resource in the UK, and one of the very best in the world. It has comprehensive news coverage and opinion from heavyweights such as George Monbiot and Nicholas Stern. The overall environmental site has separate home pages for major topics such as climate change and sub-topics such as carbon emissions, coal and the IPCC. And the Guardian has recently developed the Guardian Environmental Network, which brings in reports from a wide range of other sources as well.

Climate Ark

Web address:
  www.climateark.org

To a visitor who doesn't spend a lot of time on climate-related sites, this is in some ways an odd blog, with "biocentric commentary". The star of the site is the news feed, which draws on a wide range of sources – and, when you want to delve into a subject more deeply, customised search. The news feed is amazing – global in reach, extensive in depth, and full of interesting and timely articles. Climate Progress, featured above, is far from the only site to feature a news feed from Climate Ark, and rightly so.

The Story of Stuff

Web address:
  www.storyofstuff.com

The Story of Stuff is a 21-minute movie that ties the consumerism rampant in the industrialised world to its environmental impacts in an entertaining and informative way. It's fully viewable online and has been seen millions of times. A great discussion starter and good for classroom use. It has been criticised as one-sided, and is, in the sense that it has a point of view. But any teenager is likely to be capable of disentangling the facts presented from the opinion and participating in a useful discussion of each.

Watts Up with That
Web address:
  www.wattsupwiththat.com

An articulate blog, but firmly in the climate change sceptics camp. For instance, a recent headline was: "Climate change not all man-made". This may be true, but is it important, when the man-made kind is upending the conditions of human (and other) life on Earth? Watts Up with That recently attacked James Hansen as "no longer a scientist" – but then approvingly reprinted his attack on cap and trade. (Hansen prefers a carbon tax; Watts Up with That would prefer neither.) Run by Anthony Watts, a retired weatherman – weathermen seem to be prominent in climate change sceptic and denier circles – this blog is frustrating to read, but a good presentation of what the "other side" is thinking on climate change.

Wikipedia
Web address:
  www.wikipedia.org

Wikipedia has taken a lot of criticism recently because its policy of allowing edits by users sometimes results in misinformation being posted and remaining on the site for various periods of time. However, it's generally accurate, and an indispensable resource for getting a quick orientation to a subject – such as to the many topics related to climate change. (Wikipedia's tendency to be quite technical is more of a problem here than its occasional one-sidedness.) You can then follow up with deeper research in more authoritative sources, many of which will be listed at the end of the Wikipedia article itself.

Get Rid of Warming Blog
Web address:
  www.getridofwarming.org

Get Rid of Warming, or GROW, is the site supporting this book. It's also a blog focussed on efforts to halt global warming and climate change. Its motto is: "Saving the world, one carbon sink at a time". The GROW blog features news and references for those who understand that climate change must not only be slowed, but stopped and reversed. Part of its purpose is to develop an ever-improving answer to a question critical to humanity: are we truly in runaway climate change? The blog also cites key news articles and commentary on climate change and humanity's response to it, along with related environmental trends.

Appendix D. Winners and Losers

The ability of any consumer today to "go green" is impaired by the infrastructure available. To work to cut one's carbon emissions, while worthwhile, is difficult, expensive and only partially effective. Even some determined activists report only being able to manage cuts of about 50% in emissions.

A key part of going green is to favour "green" companies and organisations with one's spending, choice of school, employment, where to live and so on. Even a small "green premium" from consumers will drive many billions in new investment from companies, governments and non-profits – and encourage them to pressure their suppliers, stakeholders and citizens to do more as well.

Favouring "green" products and services will give everyone more, easier, and less expensive choices in cutting their personal carbon emissions, creating a virtuous cycle of reduced emissions.

We should soon each – as individuals and as organisations – be able to establish our own "green supply chain", in which we increasingly buy from, and sell to, "green" suppliers and customers. Eventually, it should be easier to live "green" than to do anything else.

What follows is a partial and idiosyncratic list of organisations that have been "early adopters" in going green and that have, in most cases, publicised their efforts as well – those that, to my understanding, have the sizzle as well as the steak.

This list is of necessity the barest starting point. I urge you to look for broader lists, local lists, lists customised for your specific interests and profession and so on. Use them to guide your personal and, with your organisation's consent, business buying decisions.

This list is also heavily British for this UK-centric edition of this book. For consumer-facing business, the UK is also where much of the action is, compared to the US at least.

Finally, a word about "air miles": the carbon footprint for shipping a specific product by land and sea is small, and it's often impossible to know how far and by what modes raw materials and components have travelled. So I don't worry much about shipping distances for most products, with two exceptions: food and drink, of which we each consume many pounds a week, and known, egregious examples of companies filling planes or ships with heavy, low-value products. (Bottled water from Fiji and cheap wines from Australia being, for me at my home base in London, examples of both.) I favour companies that avoid the worst practices. My standards and, I'm sure, yours will rise as practices and available information improve.

Disclosure: I've consulted for HSBC, one of the organisations listed below, and may do so for others I recommend in the future. I have, of course, tried to keep such relationships from affecting the list.

Winners: Places

I suggest favouring the following places – and others that "go green" – when deciding where to live, where to go to school, where to invest, where to visit and where to hold business meetings and events, all of which build on each other in a virtuous cycle. For even a few people to make such choices will incite first jealousy, then the competitive urge, in other localities.

As environmental problems worsen, those in "green" places are likely to be better protected, happier and healthier than people elsewhere. This will attract the "best and brightest" – and greenest – new residents and visitors in another virtuous cycle. So, if you have a voice in planning for a locality, get in early and beat the rush!

The next step in "going green" for localities is the Transition Towns movement, described below, or similarly broad and deep local sustainability efforts. Read up on Transition Towns – and participate if you can – to be up to speed with the cutting edge.

Favour places that are making an effort – and let people know you're doing it, especially people in the places where you spend your money. That way the "green premium" will follow you wherever you go.

Germany

Germany may be the greenest major country in the world. With the geographic deck stacked against them in terms of easy access to green power, they've done very well, pioneering solar power, solar water heating, wind power, passivhausen that can be kept warm by body heat – in a German winter! – and more. And they've done this through reasonably sized but consistent public investments and preferential rates for consumer-generated power – little drama, just impressive progress and new models other countries can follow today.

As mentioned previously, Germany – not even in the top 10 countries by population – has recently lead the world in renewable energy investments. Of course, having a large, well-organised and successful Green party helps. But that isn't an accident, and the success of their Greens reflects well on Germany as a whole.

All while my own native country, the US, has emitted massively and acted, until recently, to undermine both the science and the politics of climate change; and while my adopted home, the UK, has spoken out boldly, but mostly left innovation in public policy to the young protesters putting themselves on the flight line to stop airport expansion.

If we as a species get out of this mess in one piece, Germany and its people will be a good part of the reason. It's really important that a major world power and heavily industrialised country is taking these steps, as it sets an example for the other industrialised countries that have emitted most of the greenhouse gases now in the atmosphere, as well as the industrialising ones that are increasing their own share.

So buy German products and services. If you're in the UK or Europe, take a train to Germany for your holiday. Buy a German a drink – or dinner. Learn German. Flirt with Germans. And tell your vendor, hotelier, language instructor, new boyfriend or girlfriend etc. why you're doing it.

Among other industrialised countries, Denmark and Japan are two that took steps after the Arab oil embargo of the 1970s to cut fossil fuel usage and therefore, as a side effect, emissions. Denmark in particular is very low in its greenhouse gas emissions per unit of economic productivity. If only then-President Jimmy Carter – ironically, a nuclear engineer by training – had done this in the US instead of whinging away in that darned sweater!

Their previous efforts make it hard, ironically, for the Danes and Japanese to make the percentage cuts called for by Kyoto and other agreements today – they're cutting into muscle rather than fat. So two cheers for them, but three for Germany.

Costa Rica

Costa Rica has made amazing strides on environmentalism and against climate change. Realising its dependence on tourism and agriculture – and the dependence of both on the environment – Costa Rica began taking significant steps in the 1990s. This notably included putting all major environmentally related areas – energy, the environment, mines and water – under one Cabinet minister.

More than 90% of Costa Rica's energy comes from hydro-electric power, wind and geo-thermal power. Even more amazingly, when Costa Rica discovered oil several years ago, it decided to leave it in the ground rather than profit.

Environmental services are accounted for – those who damage them must pay, and local people are paid for preserving healthy forests and clear water. This is a major source of income for poor – in some cases, now formerly poor – people, and has played a direct role in the doubling of forested area in Costa Rica over the last twenty years.

Costa Rica has set the goal of becoming the world's first carbon neutral country by 2021. Even having this goal in sight is an amazing example of what a determined people can achieve.

In support of their efforts, if you're in the Western Hemisphere and are going to take a flight for a vacation, or a learning trip to find out new things you can apply at home, or a combination, Costa Rica has to be at the top of your list. Costa Rican exports – mostly sent to the US and China – to look out for include electronic components, clothing, bananas, coffee (including Fairtrade coffee) and artisan-carved wood products.

California

What was once called "the late, great state of California" is again leading social innovation, this time on climate change.

With the biggest population of any US state and a powerful economy that would put it in the Top 10 list among countries worldwide, California is a force to be reckoned with. Under famously Humvee-driving and stogie-smoking Governor Arnold Schwarzenegger – called "the Governator" after his leading role in the Terminator movies – California has led the US in mandating renewable energy, in promoting recycling and, most dramatically, in seeking higher mileage standards for cars.

Blocked by the Bush administration from imposing higher-than-national standards, as it did some decades ago in the battle against smog, Schwarzenegger looks set to get along better with new President Obama than he did with his fellow Republican, George W Bush – though he may now have to work harder to keep California surfing a much bigger and faster-breaking wave.

London

London is not thoroughly green in the manner of Germany, or doing as much as California, but may well be the easiest world-class city in which to go without a car. Its moderate weather also helps reduce heating expenses and emissions, and – aided by the famous British stiff upper lip – helps keep air conditioning rarer than in many other places.

The London Underground – famously called the Tube – is the first and most extensive system of its kind in the world. It gets a lot of stick, as Brits say, but try getting across, say, LA in 30 minutes during rush hour before you criticise the Tube.

The Tube is increasingly well-integrated with trains coming into London and with the ever-more-robust bus network, with most trips including a good bit of that old environmentalist staple, walking. (My daughter, wearing a pedometer at the tail end of rehabilitating an injured hip, once walked 10 miles in a normal day of errands across London.)

If the British Government's commitment to keeping London a major air transport hub bothers you, as it does me, then fly Virgin, by some measures the greenest airline, and avoid Ryanair, a budget airline which publicly trashes climate change as a problem. Offset your air miles and consider donating to Plane Stupid, a particularly effective anti-airport-expansion group run by young people.

Ken Livingstone, formerly called "Red Ken", was mayor of London for eight productive years. He made a landmark commitment for London to fight climate change very early on – in 2001 – and took steps to implement it. The piece de resistance for London's relatively green transport network is the congestion charge, an £8 daily toll for driving in the centre of the city. Introduced by Livingstone in 2003, the "C Charge" raises money for mass transit, reduces traffic, keeps the city's streets driveable and makes room for buses to run on time. It's no accident that London narrowly won the 2012 Olympic Games award shortly after the C Charge was introduced.

After a closely fought election, tousle-haired Tory Boris Johnson replaced Livingstone in 2008 – with one of his first acts being to reverse a controversial extension of the congestion charging zone. He rides a bike nearly everywhere and has proposed broad support for electric cars. As a Tory, he helped convince the party to oppose expansion of Heathrow. And London was the first big city in the world to have a Transition Town (see below), Brixton. London looks set to continue its status as a leading green city.

Transition Towns

One of the most agonizing problems faced by everyone concerned about climate change is how to rally public support for the difficult steps needed to make a difference. Rob Hopkins of Totnes, in Devon in the southwest of England, is the founder and leader of a new movement called Transition Towns that is perhaps the most promising such effort in the world.

Appendix D. Winners and Losers   221

The premise of the Transition Towns movement is that Peak Oil and climate change will make a huge – and potentially disastrous – impact on how people live. The Transition Towns effort makes this into a positive by emphasising the constructive steps people can take to prepare. "Re-skilling", for instance, means re-learning DIY, mending and other skills that have been largely lost in the last few decades, all in pursuit of local self-sufficiency.

Transition Towns organisations have been founded in hundreds of towns and districts within cities. They work with governments and existing organisations, usually finding a ready hearing. If you want to take constructive action, joining or starting up a Transition Town group is a very promising way to quickly make a difference and to learn from people who are making a big difference.

Winners: Businesses

Using a "green" buying strategy is one of the most effective ways to fight climate change. Consider paying as much as, say, a 10% premium for the "green" alternative – and tell people why you're buying it. As with geographical decisions, sustained "green" consumer buying power will influence additional tens, then hundreds of billions in stock market capitalisation and investment decisions.

Marks & Spencer

Marks & Spencer, the large and well-regarded mid-range British retailer, sells everything from knickers to dresses to housewares to food. ("It's not just food, it's M&S food", says the ad tagline.) So going green would seem to be particularly difficult for them.

It has been difficult, but it's a task they've taken on cheerfully and effectively. The company's Plan A – "because there is no Plan B" – encompasses all of the many activities the company participates in. The company has cut its carbon and environmental footprint hugely, and greatly raised public awareness as well.

If you're reading this book in the UK, and can afford their sometimes high prices (within your "green allowance" of an extra 10% or so), try to increase M&S' share of your spending. (Their broad range of goods and store formats can make it a bit confusing; here in London, an M&S "Simply Food" store in Fulham has a display window full of ladies' undergarments, while the shop in Marylebone train station has little save ready meals for commuters.)

If you prefer online shopping, are pressed for time or don't have a store, or a store of the type and with the stock you need, nearby, the Marks & Spencer Web site (at http://www.marksandspencer.com) is a good place to shop. (Online shopping, by aggregating supply and delivery, is in most cases more "green" than going out to the shops yourself; and the more people do it, the greener it gets.)

The Co-operative

The Co-op is a British retailer, largely of groceries but also of many other services, that strives to be ethical in all its policies, from packaging (less) to parking (don't expect much) to pricing (low, and with profits used for socially constructive purposes). Though The Co-op doesn't blow its own horn much, independent surveys have put it at the very top of UK retailers for its green accomplishments.

The inside of a Co-op shop feels a bit homespun – but they meet the highest operational as well as ethical standards. Seek them out as often as you can, and consider becoming a member.

HSBC

Founded as The Hong Kong and Shanghai Banking Company nearly 150 years ago, HSBC is a big bank in the UK and prints money in Hong Kong. (The joke has been made that the company's outsize profits in Asia come from keeping a percentage of each such printing for itself.) In other countries, HSBC specialises in the related areas of international banking, business banking and banking for the wealthy and aspirational.

Even in the credit crunch, HSBC has largely gone from strength to strength, maintaining much of its market value while rivals such as Citigroup have crashed. This has left HSBC among the world's top few banking groups.

This perhaps surprising recent period of dominance has coincided with HSBC's becoming the world's first carbon neutral bank. Always conservative, if not downright stingy, in its spending, the company has sharply reduced its carbon emissions – finding, as have many other organisations, that it saved a great deal of money in the process – and offsets the rest. HSBC also has innovative low-carbon buildings in several cities, such as in Mexico City, and is a leader in lower-waste online banking.

HSBC invested a huge $100 million in its Climate Change Partnership with four environmental nonprofit organisations and has followed up with something quite hard for a bank to do: changes, albeit incremental ones, to its lending and investment policies.

The bank's Climate Change Champions programme, which is training thousands of employees – many tipped as future leaders of the business – in the basics of climate change and how to fight it, is a model for every company, government and business school in the world.

I've heard the opinion expressed that HSBC's efforts are "greenwash" – a thin layer of PR over unchanged corporate priorities and practices. Having worked "inside" briefly, I disagree, I think HSBC is about as far out in front of its shareholders and customers as it can afford to be, and that the challenge is with competitors to do anywhere near as much. In the meantime, patronise HSBC, especially their online options (which are also priced lower) if you can.

Losers

I think it's still early enough in people's understanding of how serious climate change is that boycotts and other punitive steps are more likely to cause bad feeling and potentially harmful disruption than to lead to improvements.

People make up organisations and populate countries, and are behind every product and service offered. People are also easily offended, especially if they feel they're being treated unfairly, and have long memories. So better, at this stage, a steady shift toward the best than an inherently uneven attempt to "name and shame" the worst.

Just for one example that's close to home for me, I can certainly see the green rationale for boycotting particularly damaging, or even all, American products and services during the destructive and inept George W Bush presidency. But such a boycott would have offended just about all Americans and hurt, not helped, efforts to turn the US around on the issue – as is happening now under President Obama.

I think punitive actions should be in support of local and national laws and international agreements, in which case they aren't potentially rash actions led by a few, but supports to the expressed and codified opinion of the world's peoples and their leaders.

There are three exceptions where the damage being caused is so great, and the stakes are so high, that action must begin soon – though, to paraphrase Lincoln, with malice toward none, and charity towards all of the people who will suffer wrenching changes as a result. They are coal-fired power plants, air travel, and products that promote deforestation, including beef and first-generation biofuels.

Deforestation

The world needs to stop deforestation and begin reforestation as a big step toward emissions reduction, so as a first step, deforestation must be stopped.

World meat and dairy consumption must be capped, as feeding precious grain to livestock is a wasteful way to use the water and land commitment tied up in the grain represents. As livestock grazes in pastures less and fattens in feedlots more, it also needs protein supplements in its food – which can be ground-up livestock remains (the source of mad cow disease) or fish meal that could otherwise feed people directly. All of which, as we face a looming food crisis, borders on the insane. (Which is what you will become as the prions of mad cow disease begin to multiply in your brain.)

Livestock – especially cattle – also emit great amounts of greenhouse gases, particularly methane. Without a cap, more arable land and fresh water than we will ever find will be needed to support growth in the number of heavy meat eaters from 1 billion today to 4 billion in a few decades. Meaning, we'll end up with limits anyway; whether through a gradual reduction or via a series of supply crises, and whether before or after destroying what's left of the environment, are the only questions.

So we need contraction and convergence for meat and dairy consumption as we do for greenhouse gas emissions – not necessarily to zero, in this case, but certainly no increases from today. This means steady reductions in Western countries to allow some growth of consumption in developing countries. Perhaps a small allotment can be provided at a low cost per person, with the market sorting out the rest. (Rationed meat is a grim prospect, but better than the alternatives.)

Other products that promote deforestation – including, ironically, first-generation biofuels – must be identified and put out of use as well. Reforestation efforts must be supported by consumer, business and government purchasing bias toward products, companies and countries which do the most.

Coal

The next offender too serious to ignore is coal. Coal accounts for about a third of the greenhouse gases going into the air, and that share is due to increase with time. Coal is also highly polluting, and coal soot is particularly damaging to snow cover and glaciers that provide water, in the form of spring and summer runoff, to much of the world's farms and people.

People badly need power, and power plants of all sorts take years to plan and build. My suggestion is that coal plants continue to be licensed to 2012 without protest. But that 2013 then be the first unlucky year for coal – of many – and that peaceful protests, boycotts, laws and trade agreements be used to stop new coal plants coming online anywhere in the world from that year on.

Soon after, countries that use coal will have to show plans for decommissioning their plants, or suffer economically as a result.

Informed sources seem to increasingly agree that "clean coal" is a dangerous fantasy, so only plants with proven emissions-neutralizing technology – which does not exist yet, and may never exist   built in from the start could be excepted.

All this is unfair to China in particular, which has huge reserves, gets three-fourths of its electricity from coal, and needs low-cost power to grow – but which also desperately needs to stop warming, desertification and glacier loss. No exceptions then.

Air travel and air freight

The third worst offender is aviation. How much you fly – and whether in economy, business class (roughly twice the impact) or first class (double the impact again) – swamps other actions you may take with regard to your carbon footprint. Each hour airborne (in economy) causes emissions of nearly a quarter ton of CO_2; for a European, every four hours in the air is worth a month's other emissions.

Though only about four percent of greenhouse gas emissions today, aircraft are uniquely resistant to reduction efforts – they must have light, portable fuel. This, plus the continuing increase in air travel and air freight, mean that aviation is projected to quadruple its share of greenhouse gas emissions in the next few decades.

Also, planes expel emissions high up in thin air, causing much greater damage. We know high-altitude emissions are the worst, but the jury is still out on just how much worse.

I think the solution for the next decade – the teens – is to prevent the growth of air travel by preventing the growth of its infrastructure. So no new airports and runways. This will force prices up, making investment in alternatives ever more attractive. The UK's Conservative party has already proposed ameliorating the impact of cost increases by offering one low-cost long trip per person per year.

That, plus consumer favouritism toward greener aircraft and toward non-air-freighted goods, will get air travel emissions under control with the least possible economic harm, while providing the rationale for new investment in cross-country and transcontinental rail service. These, in turn, will enable the reduction of air travel to a limited, over the oceans service.

This is, again, unfair to developing countries, who are well behind the West in aviation infrastructure. But they're the ones first in line to suffer from, and least able to afford the effects of, climate change. (And in the best position to steer investment toward truly effective mass transit-based development.) So, again, no exceptions.

Appendix E. Discussion questions

A book of this type is necessarily general and prescriptive. Discussing the issues in a group really brings the topic to life. Research shows that discussing climate change topics in a group is likely to lead to better decision-making than by individuals acting alone .

As a test of part of the content of this book, I led a presentation and discussion of climate change and sustainability at the Project Management Institute chapter in London in early April, 2009.

The presentation was well received, but the most exciting part of the meeting, for all concerned, was the discussion session.

The group split up into six tables of eight to nine project managers per table and discussed a range of topics relevant to the impact of climate change and sustainability on the profession.

Each table then shared its main discussion points and conclusions with the group, with everyone learning from each other. (In the online world, this is called "user-generated content" and "social networking".) The event was highly rated, with several attendees calling it the best such meeting ever.

The following questions can be used, "as is", as a starting point for a meeting or a book club discussion. They can also be used as a jumping-off point for creating your own set of discussion questions.

For convenient photocopying, the questions begin on the next page. (Copying of the questions may well be allowed under the "educational use" clause of the copyright laws.) You may also be able to find a soft copy of the questions, which you can use "as is" or modify, on my blog at getridofwarming.org.

1. How aware are you of the megatrends that predate climate change, such as population growth, deforestation and overuse of fresh water?

2. What priority should addressing these issues have vs. working to reduce greenhouse gas emissions?

3. How big a problem do you think climate change is – choose one: the #1 problem we face this century; on a par with other top problems such as poverty, hunger and terrorism; or secondary to such problems?

4. What movies, TV shows, books and press coverage have you seen about climate change? What did they convince you – or fail to convince you – of?

5. Is it credible that the conclusions reflected by the mainstream consensus on climate change – especially that we can allow 2°C of warming before climate change goes runaway – might have been overoptimistic?

6. What are some steps that could be taken to reduce emissions in the developed world? How can those steps be agreed and enforced? How should the developed world encourage, or try to require, the developing world to follow along?

7. What are some steps that could be taken to reduce emissions in the developing world? How can these steps be agreed and enforced? How much should the developed world contribute to the developing world's costs and to its need for new technologies?

8. The idea that carbon sinks have already tipped is a new one. Does it seem sensible? If so, what can be done to counteract tipping? If not, what evidence will be needed to prove they have tipped?

9. How much should the response to climate change be driven from the top down, by laws and international agreements? How much from the bottom up, by individual initiative and citizen pressure?

10. Describe a scenario in which climate change is successfully arrested and the damage reversed. What are the key two or three drivers of this scenario?

11. What specific opportunities does your country, region, state or city have to contribute to the fight against runaway climate change? To protect itself from the effects?

12. What specific opportunities does your profession, industry or other group have to contribute to the fight against runaway climate change? To protect itself from the effects?

Index

2007 Consensus, vi, xv, 41-44, 53-57, 61-63, 76, 87, 93, 100-102, 113, 144-148, 170, 180, 188, 192, 198, 205, 208, 213

acidification of seas, 9, 20-21, 34, 73, 131, 135, 141-142, 147, 151, 165, 167, 197, 200, 202, 207
Africa, 3, 129-130, 182
air conditioning, 58-59, 67, 82, 86, 89, 220
airport, 74, 183, 219, 221, 225
Al Gore, vi, x-xi, 1, 36, 41-46, 51-52, 54-55, 63, 85, 148, 157, 162, 171, 208-210
Alaska, 91, 103, 116, 186
Amazon, 105, 114, 117-118, 166
Antarctica, 108, 139
aqueduct, 184, 191
aquifer, 2, 9, 13, 15-16, 89, 131-132, 190
arable land, x, 15, 133, 162, 184, 191, 226
Arctic, xiv, 27, 84, 91, 93, 101, 103-105, 108, 117, 167, 186
Arctic Directorate, 91
Arizona, 160, 187, 190
Asia, 17, 89, 129, 133, 136-137, 155, 169, 223
Atmospheric Brown Clouds (ABCs), 17, 136-137
Australia, 16, 21, 36-39, 55, 61, 71, 177, 196, 218
automobile, ix, 17, 23, 35, 37, 64, 70-72, 81, 88, 133, 158, 162-163, 220
aviation, 225

Bangladesh, 90, 189
Beddington, John, 140
biodiversity, 166
biofuels, first-generation, 139, 199, 224, 226
biopharma, 133, 166, 204
Brazil, 90
Britain, 16, 18, 44, 55-57, 83, 175, 180-186, 210, 213-214, 217, 220-223, 235
Brown, Gordon, 44, 71, 188
Brown, Lester, 11, 212
Bush, George W, 35-36, 38, 51, 83, 89, 91, 144, 146, 179, 186, 188, 220, 224

California, xiv, 12, 54, 71, 187, 220
Canada, vi, xiv, 16, 28, 81, 90-92, 103, 116, 139, 160, 167, 189, 191, 193
carbon sinks, vii, xii, 6, 9, 25, 34, 40-43, 49, 55, 57, 60-61, 70, 73, 75-76, 78, 93-99, 101-103, 105, 107, 109, 111-115, 117, 119, 123-125, 127, 138-139, 141, 145, 147, 155, 165, 168, 195, 207, 216, 228
Carson, Rachel, 138
Carter, James Earl (Jimmy), 84, 219
Catholic, 130
CFCs, 9, 35-36, 150, 196
China, vi, xi-xii, xiv, 3, 9, 11, 13, 17, 18, 32, 37-39, 44, 46, 49, 52, 61-62, 64, 66-67, 71, 73-74, 81, 84-89, 90, 92, 106, 128, 131-132, 136, 155-156, 158, 162, 175, 178, 187, 210-211, 220, 225
Churchill, Winston, 26, 85, 175
clathrates, 98-99, 104-105, 109, 120, 122-123, 167
Climate Ark blog, 215
Climate Progress blog, 214
coal, 10, 18, 37, 57, 59, 62, 66-67, 76, 78-79, 82, 86, 150, 155-158, 160, 181-183, 196, 200, 207, 215, 224-225
coastline, 14-16, 148, 178, 180-181
Colorado River, 12, 105, 187
Columbus, 18
Congress, 35, 43, 51, 54, 85, 188
Co-operative market, 223
coral, 135, 167
Costa Rica, 219-220
cropland, 2, 10, 12-13, 60, 83, 105, 132, 133-134, 138, 204-205
cryosphere, 14, 61, 93, 96, 196, 200, 207

deforestation, vi, xii-xiv, 1-2, 9-10-11, 17-18, 20, 34, 41, 57, 59-61, 63, 68, 70, 75, 77, 93-95, 105-107, 113, 115, 122, 125-126, 137, 140, 146-147, 150, 155, 157-158, 163, 166, 173, 197, 202, 207, 212, 224, 226, 228
Democratic Party, US, 35, 54, 85, 182-183, 188, 193
Denmark, 83, 91-92, 185, 219
desertification, xiv, 6-7, 10-11, 42, 48, 116, 128, 130, 133, 178, 189, 208, 225
Dickens, Charles, 159, 190
Dongtan zero-carbon city plan, 66
drought, 10-12, 21, 36, 98, 176, 187, 189, 196

earthquakes, 89
Egypt, 177
emissions, vi, viii, xi-xv, 5-6, 10, 17, 18, 23, 31-36, 38-39, 40-42, 44, 46-55, 57-79, 81-90, 94-95, 99, 101-106, 109, 112-120, 122-127, 131, 135, 137-141, 143, 145-

152, 155, 158-167, 169, 173, 178, 180, 185, 191-193, 195, 197, 199-202, 208, 210-212, 215, 217, 219-220, 223, 225-226, 228
England, 31, 167, 184-185, 221, 235
enthalpy, vi, 34-35
Europe, vi, xii, xiv, 2, 6, 8, 10, 16-18, 31-32, 36-39, 43-44, 61-63, 70-74, 81-84, 86, 88, 90, 131, 136, 160, 162, 165, 176, 181, 183-186, 192-193, 202, 205, 219
ExxonMobil, 144

fertiliser, 9, 32, 131, 133-134, 138, 159, 204
flooding, 128, 139, 176, 180-181, 183-184, 186
Florida, 50, 165, 189
fossil fuels, 31, 37, 46-47, 58-59, 67, 89, 94, 97, 99, 106, 131, 156, 158, 160-161, 164, 178, 200-201, 219
France, 14
frankenfish, 205
FSB, 91

Gaia, vi, 7, 41, 55-56
Galileo, 196
genetically modified foods, 204-205
geo-engineering, 78, 92, 113-114, 126, 147-148, 152, 173, 197, 200, 202, 208, 214
Germany, 83, 162, 181, 183, 185, 218, 219-220
Götterdämmerung, 19
Greece, 176
Green Party, 219
Greenland, 5, 91, 96, 107, 119, 207
Guardian newspaper, 215

Hadley Centre, 125, 180, 185
Hansen, James, vi, 41, 43-46, 51-52, 63, 84, 216
Heal, Geoffrey, 68
Hetch Hetchy reservoir, 15
HIDE (Hot, Ice-free, De-evolved, Empty), 141
Himalayan glaciers, 13, 89, 96, 155
Homo Sapiens, 3
Hoover Dam, 12, 187
Hopkins, Rob, 221
Hot Earth, 141
HSBC, 199, 223-224
Hurricane Katrina, 36, 177, 186-187
Ice Ages, 2, 5, 6, 10, 29, 149
India, xii, 3, 13, 17, 44, 49, 62, 64, 66-67, 71, 74, 81, 88-89, 90, 131-132, 136, 155, 157, 162, 187
Indonesia, 60, 90, 106
Industrial Revolution, 20, 57, 65, 81-82, 129, 132
industrialisation, viii, 5, 8, 15-18, 50, 65, 90, 152, 207

industrialised countries, 3, 8-9, 11, 15, 31-32, 61, 66-67, 81, 136, 166, 215, 219
Institution of Mechanical Engineers (IMechE), vi, 75-76, 214
insurance, 14, 72, 140, 190
Intergovernmental Panel on Climate Change (IPCC), i, vi, viii, xi, 32, 35, 39, 41, 43-52, 54-55, 58, 72, 75-77, 93, 107, 124-125, 139, 144, 148, 153-155, 170, 171, 174, 208, 211, 215
International Standards Organisation (ISO), 156
IPCC (q.v.) Scenarios, xi, 46-50, 75, 77, 144, 153-154, 211
Iran, 130

Japan, 16, 39, 61, 63, 71, 84, 202, 219
Jasons, 84
Jurassic era, xiii, 4, 6, 203-204

KGB, 91
Khan, Genghis, 9, 106
Kyoto Protocol, 83

Labour Party, UK, 183, 188
Livingstone, Ken, 221
Lomborg, Bjorn, 209, 212-213
London, iii-iv, 48, 50, 75, 136, 139, 162, 165, 180-181, 184, 218, 220-222, 227, 234-235
Los Angeles, 15, 136, 221
Lovelock, James, 41, 55-56, 139
Lynas, Mark, 202, 209, 212

Machiavelli, 143
Manhattan, 189
Marks and Spencer, 54, 199, 222
McCain, John, 189
McGuire, Bill, 212
Mediterranean Sea, 176
megatrends, 1-2, 16-18, 20-21, 127-228
Mein Kampf, 26
Melbourne, 21
Mexico, 16, 189, 191, 223
Meyer, Aubrey, 149, 151
Millennium Development Goals, 20
mini-submarine, 91
Mitchell, Mukti, 213
Monbiot, George, 145, 163-165, 209-211, 215
monsoon, 133
Munich Re, 84
Muslim, 130

NASA, 41, 43, 51, 84
National Audubon Society, 19
Neanderthals, 3
Nevada, 12, 187, 190
New Orleans, 16, 36, 151, 177, 186-187, 189
New York City, 165, 186
New Zealand, 16, 178, 208

232   *Runaway*

Nobel Peace Prize, 33, 44, 53-55, 77, 153
North Pole, vii-viii, xii, 12, 28, 30-31, 42, 44, 46, 61, 84, 90-91, 94, 96, 99-104, 108, 112-113, 115-116, 119-120, 122, 124, 131, 152, 160, 168-169, 180-182, 184, 186, 195-196, 207-208

Obama, Barack, 38, 54, 62-63, 85, 127, 149, 179, 188, 189, 191-193, 205-206, 220, 224
OPEC, 71, 89

Pachauri, Rajendra, 154
Pacific Ocean, 202
Palin, Sarah, 91
Pangaea, 3, 5
passivhaus, 218
Pax Americana, 191
Peak Wood, 10
Pearl Harbor, 144-145, 199
permafrost, 27, 94, 98-99, 101, 103-105, 116-117, 122-123, 168, 195
Peru, 13, 96
Petri dish, 7-8, 11, 13
pH (acidity), vii, 135, 151-152, 165, 207

Quakers, 55

rainforest, 105, 106, 117-118, 166, 207
Reagan, Ronald, 35, 89
Real Climate blog, 214
recycling, 220
Republican Party, US, 35, 183, 188-189, 201, 220
re-skilling, 222
Romm, Joe, 144-145, 209, 214
Rudd, Kevin, 36, 38
Russia, vi, xiv, 2, 17, 28, 46, 61, 81, 83, 90-92, 100, 103, 167, 176, 181-182, 193

Sahara Desert, 176
San Francisco, 15, 162, 234
Schwarzenegger, Arnold, 220
Semmelweis, Ignaz, xv
sequestration, 105
Shanghai, 66, 223
Sierra Nevada mountains, 12

Silicon Valley, 189
smog, 9, 17-18, 33, 97, 99, 112, 119, 128, 136-137, 169, 181, 220
Snowball Earth, 4-5, 142
soot, 33, 51, 69, 97-99, 137, 155, 196, 224
status quo ante (SQA), 151, 200, 207
Stern, Lord Nicholas, vi, 41, 43-44, 52-53, 63, 68, 180, 208, 211, 215
subsidence, 187
Sullivan, Andrew, 48
Supreme Court, 182

Thames River, 180
Thatcher, Margaret, 42, 57
Thermohaline Circulation, 180
thermometer, 5, 27-28, 29
tipping point, iii, vi-vii, xi-xii, 6, 17, 25, 34, 40-41, 43, 49, 55, 57, 70, 72-73, 75-76, 92-95, 99, 102-104, 107, 109, 112-119, 124, 127, 133, 138, 141, 155, 165, 195, 207-208, 223, 228
Totnes, UK, 221
Transition Towns, 218, 221-222

United Kingdom (UK), iii, vi, 31, 37, 41, 52, 54, 75, 81-82, 117, 125, 140, 149, 157, 161-162, 173, 180-185, 187-188, 190, 199, 213, 215, 217, 219, 222-223, 225
UK Climate Impact Programme, 77
United Nations, 37
United States of America (USA), i, vi, viii-xii, xiv, 6, 10-12, 14-16, 19, 31-32, 35-39, 43-44, 46, 51, 54-55, 58-59, 61-64, 66-68, 71-72, 74, 77, 81-92, 118, 127, 131, 136, 151, 156, 160-162, 165, 173, 177, 181-183, 185-193, 196, 198-199, 201-202, 205, 214, 217, 219, 220, 224
US Global Change Research Program, 77
US Congress, 51
US Congressional, 51

Vietnam, 155

Washington, DC, 50, 165
Watts Up With That blog, 216
World Wildlife Federation (WWF), 19

About the Author

Floyd Smith holds degrees from the University of San Francisco and the London School of Economics. For more than 20 years, he has written on science and technology while holding management positions in technology, financial and consulting companies. His books have sold over one million copies. Floyd is editor of getridofwarming.org.

Editorial Notes

The first draft of this book was composed on a RIM Blackberry Curve 8820, with each sub-section written as one or more separate note documents. It was then imported into Microsoft Outlook on an HP Compaq 6720s laptop running Windows 7 and revised and edited to completion in Microsoft Word 2003.

Guidance for page layout in Word and certain editorial matters was provided by *Perfect Pages* by Aaron Shepard (2007). The text is in British English, with the London Times Style Guide used for reference. (Thus no "ize" at the end of words, but "ise" instead; also, sadly, no "whilst" or "amongst".) Any errors are the responsibility of the author.